THE TRANSACTIONAL INTERPRETATION
OF QUANTUM MECHANICS

Providing a comprehensive exposition of the transactional interpretation (TI) of quantum mechanics, this book sheds new light on long-standing problems in quantum theory such as the physical meaning of the "Born Rule" for the probabilities of measurement results, and demonstrates the ability of TI to solve the measurement problem of quantum mechanics. It provides robust refutations of various objections and challenges to TI, such as Maudlin's inconsistency challenge, and explicitly extends TI into the relativistic domain, providing new insight into the basic compatibility of TI with relativity and the meaning of "virtual particles." It breaks new ground in approaches to interpreting quantum theory and presents a compelling new ontological picture of quantum reality. This substantially revised and updated second edition is ideal for researchers and graduate students interested in the philosophy of physics and the interpretation of quantum mechanics.

RUTH E. KASTNER is a research associate and member of the Foundations of Physics group at University of Maryland, College Park (UMCP). She is the recipient of two National Science Foundation awards for research in time symmetry issues and transactional interpretation, and of a 2021 Research Award from the Alumni Association of the University of Maryland.

THE TRANSACTIONAL INTERPRETATION OF QUANTUM MECHANICS

A Relativistic Treatment

Second Edition

RUTH E. KASTNER

University of Maryland

CAMBRIDGE
UNIVERSITY PRESS

CAMBRIDGE
UNIVERSITY PRESS

University Printing House, Cambridge CB2 8BS, United Kingdom

One Liberty Plaza, 20th Floor, New York, NY 10006, USA

477 Williamstown Road, Port Melbourne, VIC 3207, Australia

314–321, 3rd Floor, Plot 3, Splendor Forum, Jasola District Centre, New Delhi – 110025, India

103 Penang Road, #05–06/07, Visioncrest Commercial, Singapore 238467

Cambridge University Press is part of the University of Cambridge.

It furthers the University's mission by disseminating knowledge in the pursuit of
education, learning, and research at the highest international levels of excellence.

www.cambridge.org
Information on this title: www.cambridge.org/9781108830447
DOI: 10.1017/9781108907538

First published 2022

A catalogue record for this publication is available from the British Library.

ISBN 978-1-108-83044-7 Hardback

Contents

Preface

This book came about as a result of my profound dissatisfaction with the existing "mainstream" interpretations of quantum theory and my conviction that the unusual mathematical structure of quantum theory indeed reflects something about physical reality, however subtle or hidden. In my early days as a physics graduate student, I was a "Bohmian"; however, I became dissatisfied with that formulation for reasons discussed here and there throughout the book. It is my hope that, even if the reader does not come away convinced of the fruitfulness of the present approach, this presentation will serve as an invitation to further far-ranging and open discussion of the interpretational possibilities of quantum theory.

I have attempted to make much of the book accessible to the interested layperson with a mathematics and/or physics background, and to indicate where more technical sections can be omitted without losing track of the basic conceptual picture. For those in the field, I have endeavored to take into account as much as possible of the relevant literature and to use notes where a technical and/or esoteric point seems relevant. Chapters 5 and 6 are the most technical and may be omitted without losing track of the basic conceptual picture.

The first edition of this book presented my relativistic development of the Transactional Interpretation with a focus on ontology. Specifically, it made a case for the idea that quantum states and quantum dynamics (such as the Schrödinger equation) represent entities and processes that transcend the spacetime construct and should be understood as a form of physical possibility. Thus, I termed the proposed approach the "Possibilist Transactional Interpretation" (PTI). Since that time, I have developed in more quantitative detail the relativistic features, and the approach has become known in the literature as the "Relativistic Transactional Interpretation" (RTI). Both acronyms refer to the same model; they simply emphasize different aspects. In revising the presentation from the earlier edition, I have left the term PTI in place when the possibilist aspects are being discussed. However, it should be kept in mind that whether the "P" or the "R" appears in front of "TI," we are still dealing with the same proposal: a relativistic elaboration and extension of the original TI.

This second edition includes new material that shows how RTI completely nullifies the Maudlin objection, provides a quantitative account of the circumstances surrounding emission and absorption, supports established decoherence theory with a resolution to the problem of definite measurement results, provides a quantitative criterion for the micro/macro distinction, and provides a physical account of the emergence of an "arrow of time." The first four chapters are primarily conceptual in nature, giving an overview of the challenges posed by quantum theory and an introduction and overview of the transactional picture, including a proposed possibilist ontology. Chapters 5 and 6 include more technical material on the relativistic extension of TI and its applications. Chapter 7 discusses metaphysical issues in some detail, and Chapter 8 discusses both conceptual and quantitative aspects of the relation between spacetime and the transactional process. Chapter 9, the Epilogue, sums up key points and examines historical and cultural contexts relevant to consideration of the proposed approach.

I am grateful to many colleagues, friends, and family members who gave generously of their time and energy to critically read drafts of various chapters, to offer comments, and to discuss material appearing herein. I would like to thank the following people whose interest and intellectual generosity has supported my continued exploration of TI: Harald Atmanspacher, Stephen Brush, Jeffrey Bub, Michael Devitt, Donatello Dolce, Tim Eastman, Heidi Fearn, Shan Gao, John Hagelin, David Hestenes, Tim Hodgkinson, Michael Ibison, Klee Irwin, Brian Josephson, Menas Kafatos, Joseph Kahr, Stuart Kauffman, Olimpia Lombardi, James Malley, Doug Marmon, David Miller, Jeffrey Mishlove, John Norton, John Rather, Ross Rhodes, Steven Savitt, Andreas Schlatter, Allen Stairs, Henry Stapp, and William Unruh. Of course, any errors are my responsibility alone; and my sincere apologies to anyone whom I've inadvertently overlooked in the lists above.

Finally, I wish to thank my daughter, philosopher-artist Wendy Hagelgans, for valuable discussions concerning the nature of time and for drawing many of the images in this book. My other daughter, Janet, provided encouragement and inspiration by her example of perseverance in the face of challenge as she has pursued personal and career goals. My husband, Chuck, provided a sounding board as well as nonstop support and encouragement, as did my mother, Bernice Kastner. I would like to dedicate this book to my beloved family, including the memory of my late father Sid Kastner, a physicist who was also fascinated by our elusive reality, seen and unseen, and to my grandson Connor, whom I fervently hope will inherit a better world than the one that has been crafted by humans thus far.

It is my hope that this revised edition will answer many (hopefully, most) of the questions that arise concerning the ability of the Transactional Interpretation to address and resolve the vexing challenges that quantum theory presents to us.

1

Introduction

Quantum Peculiarities

1.1 Introduction

Richard Feynman said: "Our imagination is stretched to the utmost, not, as in fiction, to imagine things which are not really there, but just to comprehend those things which are there."[1] This book is an account of my attempt to meet this challenge, which comes to us in the form of the strange character of quantum theory. Specifically, it presents an overview and further development of the Transactional Interpretation (TI) of Quantum Mechanics, first proposed by John G. Cramer (1980, 1983, 1986, 1988). Quantum theory itself is an abstract mathematical construct that happens to yield very accurate predictions of the behavior of large collections of identically prepared microscopic systems (such as atoms). But it is just that: a piece of mathematics (together with rules for its application). The interpretational task is to understand what the mathematics signifies physically; in other words, to find "a way of thinking such that the law [i.e., the theory] is evident," as expressed in the quotation from Richard Feynman that introduces Chapter 3. Yet quantum theory has been notoriously resistant to interpretation: most "commonsense" approaches to interpreting the theory result in paradoxes and riddles. This situation has resulted in a plethora of competing interpretations, some of which actually change the theory in either small or major ways.

One rather popular approach is to suggest that quantum theory is not "complete" – that is, it lacks some component(s) that, if known, would resolve the paradoxes – and that is why it presents apparently insurmountable interpretational difficulties. Some current proposed interpretations, such as Bohm's theory, are essentially proposals for "completing" quantum theory by adding elements to it that (at least at first glance) seem to resolve some of the difficulties.

---

[1] *The Character of Physical Law*, chapter 6, "Probability and Uncertainty: The Quantum Mechanical View of Nature" (Cambridge: MIT Press, 1967), pp. 127–28.

(That particular approach will be discussed below, along with other "mainstream" interpretations.) In contrast to that view, this book explores the possibility that quantum mechanics *is* complete and that the challenge is to develop a new way of interpreting its message, even if that approach leads to a strange and completely unfamiliar metaphysical picture. Of course, strange metaphysical pictures in connection with quantum theory are nothing new: Bryce DeWitt's full-blown "many worlds interpretation" (MWI) is a prominent example that has entered the popular culture. However, I believe that TI does a better job by accounting for more of the quantum formalism and that it resolves other issues facing MWI. It also has the advantage of providing a physical account of the measurement process without injecting any ad hoc changes into the basic dynamics.

1.1.1 Quantum Theory Is About Possibility

Besides presenting the relativistic elaboration of John Cramer's original Transactional Interpretation (Cramer, 1980, 1983, 1986, 1988), this work will explore the view that quantum theory is describing an unseen world of possibility that lies beneath, or beyond, our ordinary, experienced world of actuality. Such a step may, at first glance, seem far-fetched, perhaps even an act of extravagant metaphysical speculation. Yet there is a well-established body of philosophical literature supporting the view that it is meaningful and useful to talk about possible events, and even to regard them as real. For example, the pioneering work of David Lewis made a strong case for considering possible entities as real.[2] In Lewis' approach, those entities were "possible worlds": essentially different versions of our actual world of experience, varying over many (even infinite) alternative ways that "things might have been." My approach here is somewhat less extravagant:[3] I wish to view as physically real the possible quantum *events* that might be, or might have been, actualized. So, in this approach, *those possible events are real, but not actual; they exist, but not in spacetime.* The *actual* event is the one that can be said to exist as a component of spacetime. I thus dissent from the usual identification of "physical" with "actual": an entity can be physical without being actual. In more metaphorical language, we can think of the observable portion of reality (the actualized, spacetime-located portion) as the "tip of an iceberg," with the unobservable, unactualized, but still real, portion as the submerged part (see Figure 1.1).

Another way to understand the view presented here is in terms of Plato's original dichotomy between "appearance" and "reality." His famous allegory of

---

[2] Lewis' view is known as "modal realism" or "possibilist realism."
[3] So, for example, I will not need to defend the alleged existence of "that possible fat man in the doorway" from the "slum of possibles," a criticism of the modal realist approach by Quine (1953, p. 15).

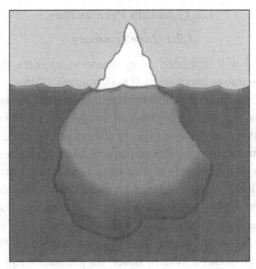

Figure 1.1 Possibilist TI: the observable world of spacetime events is the "tip of the iceberg" rooted in an unobservable manifold of possibilities transcending space-time. These physical possibilities are what are described by quantum theory. Drawing by Wendy Hagelgans.

the Cave proposed that we humans are like prisoners chained in a dark cave, watching and studying shadows flickering on a wall and thinking that those shadows are real objects. However, in reality (according to the allegory) the real objects are behind us, illuminated by a fire that casts their shadows on the wall upon which we gaze. The objects themselves are quite different from the appearances of their shadows (they are richer and more complex). While Plato thought of the "unseen" level of reality in terms of perfect forms, I propose that the reality giving rise to the "shadow" objects that we see in our spacetime "cave" consists of the quantum objects described by the mathematical forms of quantum theory. Because they are "too big," in a mathematical sense, to fit into spacetime (just as the objects casting the shadows are too big to fit on a wall in the cave, or the submerged portion of the iceberg cannot be seen above the water) – and thus cannot be fully "actualized" in the spacetime theater – we call them "possibilities." But they are *physically real* possibilities, in contrast to the way in which the term "possible" is usually used. Quantum possibilities are physically efficacious in that they *can* be actualized and thus can be experienced in the world of appearance (the empirical world).

This basic view will be further developed throughout the book. As a starting point, however, we need to take a broad overview of where we stand in the endeavor of interpreting the physical meaning of quantum theory. I begin with some notorious peculiarities of the theory.

1.2 Quantum Peculiarities

1.2.1 Indeterminacy

The first peculiarity I will consider, *indeterminacy*, requires that I first discuss a key term used in quantum mechanics (QM), namely, "observable." In ordinary classical physics, which describes macroscopic objects like baseballs and planets, it is easy to discuss the standard physical properties of objects (such as their position and momentum) as if those objects always possess determinate (i.e., well-defined, unambiguous) values. For example, in classical physics one can specify a baseball's position x and momentum p at any given time t. However, for reasons that will become clearer later on, in QM we cannot assume that the objects described by the theory – such as subatomic particles – always have such properties independently of interactions with, for example, a measuring device.[4] So, rather than talk about "properties," in QM we talk about "observables" – the things we can observe about a system based on measurements of it.

Now, applying the term "observable" to quantum objects under study seems to suggest that their nature is dependent on observation, where the latter is usually understood in an anthropocentric sense, as in observation by a conscious observer. The technical philosophical term for the idea that the nature of objects depends on how (or whether) they are perceived is "antirealism." The term "realism" denotes the opposite view: that objects have whatever properties they have independent of how (or whether) they are perceived, that is, that the real status or nature of objects does not depend on their perception.

The antirealist flavor of the term "observable" in quantum theory has led researchers of a realist persuasion – a prominent example being John S. Bell – to be highly critical of the term. Indeed, Bell rejected the term "observable" and proposed instead a realist alternative, "beable." Bell intended "beable" to denote real properties of quantum objects that are independent of whether or not they are measured (one example being Bohmian particle positions; see Section 1.3.3). The interpretation presented in this book does not make use of "beables," although it shares Bell's realist motivation: quantum theory – by virtue of its impeccable ability to make accurate predictions about the phenomena we can observe – is telling us something about reality, and it is our job to discover what that might be, no matter how strange it may seem.[5]

---

[4] The apparent "cut" between macroscopic (e.g., a measuring device) and microscopic (e.g., a subatomic particle) realms has been one of the central puzzles of quantum theory; it is also known as the "shifty split." We will see (in Chapter 3) that under the transactional interpretation this problem is solved; the demarcation between quantum and classical realms need not be arbitrary (or based on a subjectivist appeal to an observing "consciousness").

[5] The realist accounts for the success of a theory in a simple way: it describes something about reality. Antirealist and pragmatic approaches such as "instrumentalism" – that theories are just instruments to predict phenomena – can provide no explanation for why the successful theory works better than a competing theory. A typical

I will address in more detail the issue of how to understand what an "observable" is in the context of the transactional interpretation in later chapters. For now, I simply deal with the perplexing issue of indeterminacy concerning the values of observables, as in the usual account of QM.

Heisenberg's famous "uncertainty principle" (also called the "indeterminacy principle") states that, for a given quantum system, one cannot simultaneously determine physical values for pairs of incompatible observables. "Incompatible" means that the observables cannot be simultaneously measured and that the results one obtains depend on the order in which they are measured. Elementary particle theorist Joseph Sucher has a colorful way of describing this property. He observes that there is a big difference between the following two processes: (1) opening a window and sticking your head out and (2) sticking your head out and then opening the window.[6]

Mathematically, the *operators* (i.e., the formal objects representing observables) corresponding to incompatible observables do not commute;[7] that is, the results of multiplying such operators together depend on their order. Concrete examples are position, whose mathematical operator is denoted X (technically, the operator is really multiplication by position x), and momentum, whose operator is denoted P.[8] The fact that X and P do not commute can be symbolized by the statement

$$XP \neq PX.$$

Thus, quantum mechanical observables are not ordinary numbers that can be multiplied in any order with the same result; instead, you must be careful about the order in which they are multiplied.

It is important to understand that the uncertainty principle is something much stronger (and *stranger*) than the statement that we just can't physically measure, say, both position and momentum because measuring one property disturbs the other one and changes it. Rather, in a fundamental sense, the quantum object *does not have* a determinate (well-defined) value of momentum when its position is being detected, and vice versa. This aspect of quantum theory is built into the very mathematical structure of the theory, which says in precise logical terms that there simply is no yes/no answer to a question about the value of a quantum object's position when you are measuring its momentum. That is, the question "Is the particle at position x?" generally has no yes or no answer in quantum theory in the context of a momentum measurement. This is the puzzle of quantum

account in support of such approaches would say that the demand for an explanation for why the theory works simply need not be met. I view this as an evasion of a perfectly legitimate, indeed crucial, question.

[6] Comment by Professor Joseph Sucher in a 1993 UMCP quantum mechanics course.

[7] "Commute" literally means "go back and forth"; so that the standard commuting property is expressed by noting that for two ordinary numbers a and b, $ab = ba$.

[8] The mathematical form of P (in one spatial dimension) is given by $P = (h/i)(d/dx)$.

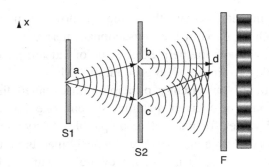

Figure 1.2 The double-slit experiment.
Source: http://en.wikipedia.org/wiki/File:Doubleslit.svg.

indeterminacy: quantum objects seem not to have precise properties independent of specific measurements designed to detect those specific properties.[9]

A particularly striking example of indeterminacy on the part of quantum objects is exhibited in the famous two-slit experiment (Figure 1.2). This experiment is often discussed in conjunction with the idea of "wave/particle duality," which is a manifestation of indeterminacy. (The experiment and its implications for quantum objects are discussed Feynman et al. (1964, chapter 1); I revisit this example in more detail in Chapter 3.)

If we shine a beam of light through two narrow slits, we will see an interference pattern (see Figure 1.2). This is because light behaves (under some circumstances) like a wave, and waves exhibit interference effects. A key revelation of quantum theory is that material objects (i.e., objects with nonzero rest mass, in contrast to light) also exhibit wave aspects. So one can do the two-slit experiment with quantum particles as well, such as electrons, and obtain interference. Such an experiment was first performed by Davisson and Germer in 1928 and was an important confirmation of Louis de Broglie's hypothesis that matter also possesses wavelike properties.[10]

The puzzling thing about the two-slit experiment performed with material particles is that it is hard to understand what is "interfering": our classical common sense tells us that electrons and other material particles are like tiny billiard balls that follow a clear trajectory through such an apparatus. In that picture, the electron must go through one slit or the other. But if one assumes that this is the case and calculates the expected pattern, the result will *not* be an interference pattern. Moreover, if one tries to "catch it in the act" by observing which slit the electron went through, this procedure will ruin the interference pattern. It turns out that

---

[9] The exception is properties belonging to a compatible observable (whose operator commutes with the one being measured). Bohmians dissent from this characterization of the theory; this will be discussed below.
[10] Davisson (1928).

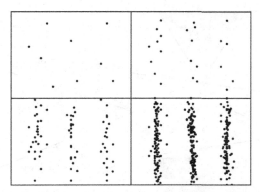

Figure 1.3 Typical results of a double-slit experiment showing the buildup of an interference pattern of single electrons.

interference is seen only when the electron is left undisturbed, so that in some sense it "goes through both slits." Note that the interference pattern can be slowly built up dot by dot, with only one particle in the apparatus at a time (see Figure 1.3). Each of those dots represents an entity that is somehow "interfering with itself" and represents a particle whose position is indeterminate – it does not have a well-defined trajectory, in contrast to our classical expectations.[11]

1.2.2 Nonlocality

The puzzle of nonlocality arises in the context of composite quantum systems, that is, systems that are composed of two or more quantum objects. The prototypical example of nonlocality is the famous Einstein–Podolsky–Rosen (EPR) paradox, first presented in a 1935 paper (Einstein et al., 1935). The paper, entitled "Can quantum-mechanical description of reality be considered complete?," attempted to demonstrate that QM could not be a complete description of reality because it failed to provide values for physical quantities that the authors assumed must exist.

Here is the EPR thought-experiment in a simplified form due to David Bohm, in terms of spin-1/2 particles such as electrons. Spin-1/2 particles have the property that, when subject to a non-uniform magnetic field along a certain spatial direction z, they can either align with the field (which is termed "up" for short) or against the field (termed "down") (such a measurement can be carried out by a Stern-Gerlach device; see Figure 1.4).

I designate the corresponding quantum states as "$|z\,\mathrm{up}\rangle$" and "$|z\,\mathrm{down}\rangle$," respectively. The notation used here is the bracket notation invented by Dirac, and

---

[11] One of the interpretations I will discuss, the Bohmian theory, does offer an account in which particles follow determinate trajectories. The price for this is a kind of nonlocality that may be difficult to reconcile with relativity, in contrast to TI.

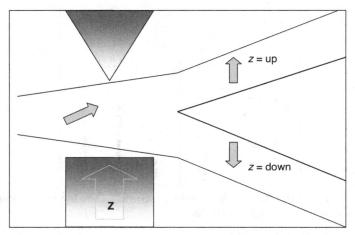

Figure 1.4 Spin "up" or "down" along the z direction in a Stern-Gerlach measurement.

the part pointing to the right is the "$|\text{ket}\rangle$." We can also have a part pointing to the left, "$\langle\text{brac}|$." (Since one is often working with the inner product form $\langle\text{brac}|\text{ket}\rangle$, the name is an apt one.) We could measure the spin and find a corresponding result of either "up" or "down" along any direction we wish, by orienting the field along a different spatial direction, say, x. The states we could then measure would be called "$|x\text{up}\rangle$" or "$|x\text{down}\rangle$," and similarly for any other chosen direction.

We also need to start with a composite system of two electrons in a special type of state, called an "entangled state." This is a state of the composite system that cannot be expressed as a simple, factorizable combination (technically a "product state") of the two electrons in determinate spin states, such as "$|x\text{up}\rangle|x\text{down}\rangle$."

If we denote the special state by $|S\rangle$, it looks like

$$|S\rangle = \frac{1}{\sqrt{2}}[|\text{up}\rangle|\text{down}\rangle - |\text{down}\rangle|\text{up}\rangle] \tag{1.1}$$

where no directions have been specified, since this state is not committed to any specific direction. That is, you could put in any direction you wish (provided you use the same "up/down – down/up" form); the state is mathematically equivalent for all directions.

Now, suppose you create this composite system at the 50-yard line of a football field and direct each of the component particles in opposite directions, say, to two observers "Alice" and "Bob" in the touchdown zones at opposite ends of the field. Alice and Bob are each equipped with a measuring apparatus that can generate a local non-uniform magnetic field along any direction of their choice (as illustrated in Figure 1.4). Suppose Alice chooses to measure her electron's spin in the z

direction. Then quantum mechanics dictates that the spin of Bob's particle, if measured along z as well, must always be found in the opposite orientation from Alice's: if Alice's electron turns out to be $|z\,\mathrm{up}\rangle$, then Bob's electron must be $|z\,\mathrm{down}\rangle$, and vice versa. The same holds for any direction chosen by Alice. Thus it seems as though Bob's particle must somehow "know" about the measurement performed by Alice and her result, even though it may be too far away for a light signal to reach in time to communicate the required outcome seen by Bob. This apparent transfer of information at a speed greater than the speed of light ($c = 3 \times 10^8$ m/s) is termed a "nonlocal influence," and this apparent conflict of quantum theory with the prohibition of signals faster than light is termed "nonlocality."[12]

Einstein termed this phenomenon "spooky action at a distance" and used it to argue that there had to be something "incomplete" about quantum theory, since, in his words, "no reasonable theory of reality should be expected to permit this."[13]

However, it turns out that we are indeed stuck with quantum mechanics as our best theory of (micro)-reality despite the fact that it does, and must, permit this, as Bell's Theorem (1964) demonstrated. Bell famously showed that no theory that incorporates local "elements of reality" of the kind presumed by Einstein can reproduce the well-corroborated predictions of quantum theory; specifically, the strong correlations inherent in the EPR experiment. *Quantum mechanics is decisively nonlocal*: the components of composite systems described by certain kinds of quantum states (such as the state (1.1)) seem to be in direct, instantaneous communication with one another, regardless of how far they may be spatially separated.[14] The interpretational challenge presented by the EPR thought-experiment combined with Bell's Theorem is that a well-corroborated theory seems to show that reality *is* indeed "unreasonable," in that it allows influences at

---

[12] I say "apparent conflict" here because it is a very subtle question as to what constitutes a genuine violation of, or conflict with, relativity. In later chapters we'll see that PTI (which is also called "relativistic TI" or RTI) can provide "peaceful coexistence" of QM with relativity, as envisioned by Shimony (2009).

[13] I am glossing over some subtleties here concerning Einstein's objection. A more detailed account of the EPR paper would note that Einstein's objection was in terms of "elements of reality" concerning the presumably determinate physical spin attributes of either electron and the fact that their quantum states seemed not to be able to specify these. As noted in the subsequent discussion, Bell's Theorem of 1964 showed that there can be no such "elements of reality."

[14] I should note that some researchers dissent from this characterization. One way out of the conclusion that quantum theory is necessarily nonlocal is to dispute the way "elements of reality" are defined. See, for example, Willem M. de Muynck's discussion at www.phys.tue.nl/ktn/Wim/qm4.htm!thermo_analogy. I am skeptical of this approach because it must introduce what appears to be an ad hoc further level of statistical randomness, beyond that of the standard theory, whose sole purpose is to enforce locality. Adherents of Everettian approaches argue that these can retain locality, but that has been disputed, for example, in Kastner (2011c), which points out that nonlocal correlations persist. This is a matter of ongoing debate in the literature. But in a nutshell, in Everettian approaches, "splitting" becomes the means to eliminate nonlocality, so the viability of "splitting" becomes crucial here, which brings us back to the fact that standard decoherence arguments do not establish measurement outcomes and therefore cannot support "splitting." Adherents of Qbism attempt to "save locality," but Henson (2015) has pointed out that their argument fails in that it also must designate explicitly nonlocal theories as "local."

apparently infinite (or at least much faster than light) speeds, despite the fact that relativity seems to say that such things are forbidden.

1.2.3 The Measurement Problem

The measurement problem is probably the most perplexing feature of quantum theory. There is a vast literature on this topic, testifying to the numerous and sustained attempts to solve this problem. Erwin Schrödinger's famous "cat" example, which I will describe below, was intended by him to be a dramatic illustration of the measurement problem (Schrödinger, 1935).

The measurement problem is related to quantum indeterminacy in the following way. Our everyday experiences of always-determinate (clearly defined, nonfuzzy) properties of objects seems inconsistent with the mathematical structure of the theory, which dictates that sometimes such properties are *not* determinate. The latter cases are expressed as superpositions of two or more clearly defined states. For example, a state of indeterminate position, let's call it "$|?\rangle$," could be represented in terms of two possible positions x and y by

$$|?\rangle = a|x\rangle + b|y\rangle \tag{1.2}$$

where a and b are two complex numbers called "amplitudes." A quantum system could undergo some preparation leaving it in this state. If we wanted to find out where the system was, we could measure its position, and, according to the orthodox way of thinking about quantum theory, its state would "collapse" into either position x or position y.[15] The idea that a system's state must "collapse" in this way upon measurement is called the "collapse postulate" (see Section 1.3.4) and is a matter of some controversy. Schrödinger's cat makes the controversy evident. I now turn to this famous thought-experiment.

Here is a brief description of the idea (with apologies to cat lovers). A cat is placed in a box containing an unstable radioactive atom which has a 50% chance of decaying (emitting a subatomic particle) within an hour. A Geiger counter, which detects such particles, is placed next to the atom. If a click is registered indicating that the atom has decayed, a hammer is released which smashes a vial of poison gas, killing the cat. Otherwise, nothing happens to the cat. With this setup, we place all ingredients in the box, close it, and wait one hour.

---

[15] The probability of ending up in x would be a^*a and in y would be b^*b. This prescription for taking the absolute square of the amplitude of the term to get the probability of the corresponding result is called the "Born Rule" after Max Born, who first proposed it. Amplitudes are therefore also referred to as "probability amplitudes." There is no way to predict which outcome will result in any individual case. TI provides a concrete, physical (as opposed to statistical or decision-theoretic) basis for the Born Rule.

The atom's state is usually written as a superposition of "undecayed" and "decayed," analogous to state (1.2):

$$|\text{atom}\rangle = \frac{1}{\sqrt{2}}[|\text{undecayed}\rangle + |\text{decayed}\rangle] \qquad (1.3)$$

Prior to our opening the box, since no measurement has been performed to "collapse" this superposition, we are (so the usual story goes[16]) obligated to include the cat's state in the superposition as follows:

$$|\text{atom} + \text{cat}\rangle = \frac{1}{\sqrt{2}}[|\text{undecayed}\rangle|\text{alive}\rangle + |\text{decayed}\rangle|\text{dead}\rangle] \qquad (1.4)$$

This superposition is assumed to persist because no "measurement" has occurred which would "collapse" the state into either alternative. So we appear to end up with a cat in a superposition of "alive" and "dead" until we open the box and see which it is, upon which the state of the entire system (atom + Geiger counter + hammer + gas vial + cat) "collapses" into a determinate result. Schrödinger's example famously illustrated his exasperation with the idea that something macroscopic like a cat seems to be forced into a bizarre superposition of alive and dead by the dictates of quantum theory, and that it is only when somebody "looks" at it that the superposed system is found to have collapsed, even though this mysterious "collapse" is never observed nor (apparently) is there any physical mechanism for it. This is the core of the measurement problem.

In less colorful language, the measurement problem consists in the fact that, given an initial quantum state for a system, quantum theory does not tell us *why* or *how* we only get one specific outcome when we perform a measurement on that system. On the contrary, the quantum formalism seems to tell us about several possible outcomes, each with a particular weight. So, for example, I could prepare a quantum system in some arbitrary state X, perform a measurement on it, and the theory would tell me that it might be A, or B, or C, but it will not tell me *which* result actually occurs, nor does it provide any reason for *why* only one of these is actually observed.

So there seems to be a very big and mysterious gap between what the theory appears to be saying (at least according to the usual understanding of it) and what our experience tells us in everyday life. We are technically sophisticated enough to create and manipulate microscopic quantum systems in the laboratory, to the extent

---

[16] TI does not have to tell the story this way; in TI one does not need to characterize the system by Equation (1.4). This fact, a major reason to choose TI over its competitors, is discussed in Chapters 3 and 4. A key component of the puzzle raised by Schrödinger's cat is that it is not at all obvious that a macroscopic object like a cat should be describable as a component of a unitarily evolving quantum state as in Equation (1.4) (indeed, I argue that it is *not*). While many current approaches recognize this issue and try to address it, I believe that TI's approach is the only noncircular and unambiguous one.

that we can identify them with a particular quantum state (such as X above). We can then put these prepared systems through various experimental situations intended to measure their properties. But, in general, for any of those measurements, the theory just gives us a weighted list of possible outcomes. And obviously, in the laboratory, we see only *one* particular outcome.

Now, the theory is still firmly corroborated in the sense that the weights give extremely accurate predictions for the *probabilities* of those outcomes when we perform the same kind of measurement on a large number of identically prepared systems (technically known as an *ensemble*). But the measurement problem consists in the fact that any individual system is still described by the theory, yet the theory fails to specify (1) what sort of interaction counts as a "measurement," (2) what that individual system's outcome will actually be, or (3) even why it has only one.

It should be emphasized that this situation is completely different from what classical physics tells us. For example, consider a coin flip. A coin is a macroscopic object that is well described by classical physics. If we knew everything about all the (classical) forces acting on the coin, and all the relevant details of the coin itself, we could in principle calculate the result of any particular coin flip. That is, we could predict with 100% certainty (or at least within experimental error) whether it would land heads or tails. But when it comes to the microscopic objects described by quantum theory, even if we start with precise knowledge of their initial states, in general the theory does not allow us to predict *any* given outcome with 100% certainty.[17] The situation is made even more perplexing by the fact that classical physics and quantum physics must be describing the same world, so they must be compatible in the limit of macroscopic objects (i.e., when the sizes of our systems become much larger than subatomic particles like electrons and neutrons). This means that macroscopic objects must also be describable (in that same limit) by quantum theory. This consideration raises the important question of: Exactly *what* is a "macroscopic object" anyway, and how is it different from the objects (like electrons) that can *only* be described by quantum theory? The quick answer, under TI, is that macroscopic objects are phenomena resulting from actualized transactions, whereas quantum objects are not. I explore this point in detail in Chapters 6–8.

Typical prevailing interpretations even encounter difficulty in specifying exactly what counts as a measurement, and (as noted above) that question is a component of the measurement problem. For example, discussions of the Schrödinger cat

---

[17] The exception, of course, is that measurements of observables commuting with the preparation observable result in determinate outcomes.

paradox have dealt not only with the bizarre notion of a cat seemingly in a quantum superposition but also with the conundrum of *when or how* measurement of the system can be considered truly finished. That is, does the observer who opens the box and looks at the cat also enter into a superposition? At what point does this superposition really "collapse" into a determinate (unambiguous) result? An example of this statement of the problem in the literature is provided by Clifton and Monton (1999):

Unfortunately, the standard dynamics [and the standard way of interpreting] quantum states together give rise to the measurement problem; they force the conclusion that a cat can be neither alive nor dead, and, worse, that a competent observer who looks upon such a cat will neither believe that the cat is alive nor believe it to be dead. The standard way out of the measurement problem is to . . . temporarily suspend the standard dynamics by invoking the collapse postulate. According to the postulate, the state vector $|\psi(t)\rangle$, representing a composite interacting "measured" and "measuring" system, stochastically [randomly] collapses, at some time t' during their interaction.... The trouble is that this is not a way out unless one can specify the physical conditions necessary and sufficient for a measurement interaction to occur; for surely "measurement" is too ambiguous a concept to be taken as primitive in a fundamental physical theory. *(p. 698)*

Thus, the measurement problem arises from the apparent unitary-only (deterministic, linear) evolution of standard quantum theory, together with ambiguity about when to invoke a non-unitary "collapse postulate" which seems not to have any physical content. The problem has recently been sharpened to an even more devastating form in the latest version of the "cat paradox" by Frauchiger and Renner (2018). These authors devised a scenario that results not just in an absurd macroscopic superposition (like Schrödinger's cat) but in an overt inconsistency: different observers will disagree on the result of a measurement that could, in principle, be performed. We will see in subsequent chapters that TI provides a very effective way out of this conundrum, including the puzzle of defining what constitutes a "measurement."

1.3 Prevailing Interpretations of QM

1.3.1 Decoherence Approaches

"Decoherence" refers to the way in which interference effects (like what we see in a two-slit experiment, Figures 1.2 and 1.3) are lost as a given quantum system interacts with its environment. Roughly speaking, decoherence amounts to the loss of the ability of the system to "interfere with itself" as the electron does in the two-slit experiment. This basic idea – that a quantum system suffers decoherence when it interacts with its environment – has been developed to a high technical degree in recent decades. In effect this research has shown that in most cases, quantum systems cannot maintain coherence, and its attendant interference effects, in processes which amplify such systems to the observable level of ordinary experience. In general, this approach to the classical level is described by a greatly increasing number of "degrees of freedom" of the system(s) under study.[18] So, decoherence shows that systems with many degrees of freedom – macroscopic systems – do not exhibit observable interference. In addition, the decoherence approach seems to provide a way to specify a determinate "pointer observable" for the apparatus used to measure a given system once the interactions of the system, apparatus, and environment are all taken into account. This apparent emergence via the decoherence process of a clearly defined, macroscopic "pointer observable" for a given measurement interaction is sometimes referred to as "quantum Darwinism," since the process seems analogous to an evolutionary process.

Many researchers have taken this as at least a partial solution to the measurement problem in that it is taken to explain why we don't see interference effects happening all around us even though matter is known to have wavelike properties. It appears to explain, for example, why Schrödinger's cat need not be thought of as exhibiting an interference pattern (which is something of a relief). But decoherence alone does not explain why the cat is clearly *either* alive *or* dead (and not in some superposition) at the end of the experiment. The reason for this is somewhat technical, and amounts to the fact that we can still have quantum superpositions without interference. Such superpositions cannot be thought of as representing only an epistemic uncertainty (uncertainty based only on lack of knowledge about something that really is determinate). In order to regain the classical world of ordinary experience, we need to be able to say that our uncertainty about the status of an object is entirely epistemic – it is just our ignorance about the object's properties – and not based on an indeterminacy inherent in the object itself. Decoherence fails to provide this. G. Bacciagaluppi emphasizes this point his entry on decoherence in the *Stanford Encyclopedia of Philosophy*:

Unfortunately, naive claims of the kind that decoherence gives a complete answer to the measurement problem are still somewhat part of the "folklore" of decoherence, and deservedly attract the wrath of physicists (e.g. Pearle 1997) and philosophers (e.g. Bub 1997, Chap. 8) alike.[19]

Here is a crude way to understand the distinction between merely epistemic uncertainty and quantum (objective) indeterminacy. Suppose I put 10 marbles in

---

[18] "Degrees of freedom" basically means "ways in which an object can move." A system of one particle (neglecting spin) can move in a spatial sense (in three possible directions), so it has three degrees of freedom. A system of three particles has nine degrees of freedom, and so on. If one assumes that the particles have spin, then additional, rotational degrees of freedom are in play.

[19] Bacciagaluppi (2016).

an opaque box; 3 red and 7 green, and then close the box. I could represent my uncertainty about the color of any particular marble I might reach in and grab by a statistical "mixture" of 30% red and 70% green. My uncertainty about those marbles is entirely contained in my ignorance about which one I will happen to touch first. There is nothing "uncertain" about the marbles themselves. Not so with a quantum system prepared in a state, say,

$$|\Psi\rangle = a \,|\, \text{red}\rangle + b \,|\, \text{green}\rangle. \tag{1.5}$$

We may be able to eliminate all interference effects from phenomena based on this object's interactions with macroscopic objects, but we have not eliminated the quantum superposition based on its state. In some sense, the state describes an *objective* uncertainty that cannot be eliminated by eliminating interference. The technical way to describe this is that the statistical state of the decohered system is a mixture, but an *improper* one. The state of the marbles was a *proper* mixture. We need a proper mixture in order to say that we have solved the measurement problem, but decoherence does not provide that.

Yet perhaps a more serious challenge for the overarching goal of the decoherence program to explain the emergence of a classical (determinate, non-interfering) realm from the quantum realm is found in the recent work of Chris Fields (2011). Fields shows that in order to determine from the quantum formalism which pointer observable "emerges" via decoherence, one must first specify the boundary between the measured system and the environment; that is, one must say which degrees of freedom belong to the system being measured and which belong to the environment. But in order to do this, one must use information available only from the macroscopic level, since it is only at that level that the distinction exists; only the experimenters know what they consider to be the system under study. So it cannot be claimed that the macroscopic level naturally "emerges" from purely quantum mechanical origins. The program is circular because it requires macroscopic phenomena as crucial inputs to obtain macroscopic phenomena as outputs.[20]

Therefore, the decoherence program does not actually solve the measurement problem, due to the persistence of improper mixtures which cannot be interpreted as mere subjective ignorance of existing ("determinate") facts or states of affairs. Nor does it succeed in the goal of demonstrating that the classical world of

---

[20] Technically, Fields' argument is independent of the scale of the phenomena; it shows that classical information must be put in to get out classical information (such as the relevant pointer observables). But in practice, this information comes from the macroscopic level – that is, the experimenters' choices concerning what they want to study. See also Butterfield (2011, p. 17) for why the decoherence program does not solve the measurement problem.

experience arises naturally from the quantum level.[21] In later chapters it will be shown that TI can readily account for the emergence of a macroscopic realm from the quantum realm. This emergence dovetails with the quantitative predictions of decoherence theory (as we will see in Section 6.5). However, in the TI account, measured systems are described by proper instead of improper mixtures; thus it achieves a resolution of the measurement problem that has eluded the standard approach regarding decoherence.

1.3.2 Many Worlds Interpretations

Many worlds interpretations are variants of an imaginative proposal by Hugh Everett (1957), which he called the "relative state interpretation." The basic core of Everett's proposal was simply to deny that any kind of "collapse" ever occurs and assert that the linear, unitary[22] evolution of quantum state vectors is the whole story. He suggested that any given observer's perceptions will be represented in one branch or other of the state vector, and that this is all that is necessary to account for our experiences. That is, the observer will become correlated with the system they are observing, and a particular outcome for the system can only be specified *relative to the corresponding state* for the observer (hence the title).

However, most researchers were not satisfied with this as a complete solution to the measurement problem. For one thing, it did not seem clear what was meant by an observer being somehow associated with many branches of the state vector. A variant proposed by Bryce DeWitt "took the bull by the horns" and asserted that these branches described actual separate worlds – that is, that the apparent mathematical evolution of the state vector into branches corresponded to an actual physical splitting of the world. This version of Everett's approach became known as the full-blown

---

[21] It should be noted that Deutsch (1999) and Zurek (2003) have presented "derivations" of the Born Rule. However, these derivations are observer-dependent, based on the specification of a non-intrinsic, classical division of objects into "system" and "observer" (or measuring device). Thus these approaches provide a subjective or purely epistemic probabilistic interpretation, based on defining ignorance on the part of some conscious observer. In contrast, TI derives the Born Rule in a physical way, with probability being a natural interpretation of what are pre-probabilistic physical weights. Thus objective probability arises out of a specific physical entity in TI – the incipient transaction. TI's physical, as opposed to epistemic, approach to probability is appropriate to the interpretation of quantum theory as being about objective, rather than subjective, probabilities. Another way to put it: Zurek's and Deutsch's approaches are *epistemic* motivations in the same way that Gleason's is a "mathematical motivation" (as characterized by Schlosshauer and Fine, 2003). Insofar as they presuppose the presence of a classical "observer," they show consistency of quantum probabilities with what such an observer would observe, rather than deriving the probabilities in terms of a physical referent. The handicap hindering such accounts is that they must work with state vectors as the only physical referent. They do not have a physical referent for the projection operators (incipient transactions) which carry the real physical content of objective probabilities in quantum theory.

[22] "Linear" means that the quantum state only appears in the first power, and "unitary" means that no physically or mathematically ambiguous "collapse" has occurred. I refer to a "state vector" rather than a "wave function" because the former is the most general mathematical form of the quantum state: an element of Hilbert space. The wave function is just an amplitude obtained from projecting the state vector into a basis.

"many worlds interpretation."[23] (Perhaps not surprisingly, the MWI has become the basis for many science fiction stories – a good example being the episode "Parallels" of *Star Trek: The Next Generation* (seventh season) in which the character Worf finds himself "transitioning" between different possible Everettian worlds with differing versions of events.) Proponents of MWI rely on decoherence in order to specify a basis for the splitting of worlds – that is, to explain why splitting seems to happen with respect to possible positions of objects rather than, say, their momenta or any other mathematically possible observable.

Other Everettians, who adhere to a version called the "bare theory," prefer not to subscribe to an actual physical splitting of worlds, but instead attribute a quantum state to an observer and describe that observer's mental state as branching. Adherents of the bare theory argue that consistency with experience is achieved by noting that a second, nonsplitting observer (call him Bob) can always ask the first observer (Alice, who is observing a quantum system) whether she sees a determinate result, and Alice can answer yes without specifying what that result is.[24] Thus, an observer's state will either split along with a previous observer (if they inquire what the particular result was) and each of their branches will be correlated in a consistent way with the first observer's branches; or it will not split, and the second observer will still receive a consistent answer, if they only ask whether the first observer perceived a determinate result (but does not ask what the specific result is).

However, Bub (1997) and Bub et al. (1997) have argued that this approach ultimately fails to solve the measurement problem. Their critique is rather technical, but it boils down to two essential observations. (1) It turns out that there is an arbitrariness about whether the first observer will report "yes" or "no" concerning the determinateness of their perceptions and that the choice of "yes" can be seen as analogous to choosing a "preferred observable" – that is, a particular observable that is assumed to always have a value. But that assumption contradicts the original intent of the interpretation – it is supposed to be a "bare" theory, after all, with no additional assumptions necessary besides the linear, unitary development of the quantum state. (2) It is not enough for Alice to simply report that she perceived a determinate result: we commonly take ourselves not only to perceive something definite, but also to perceive *what* that thing is. Bub et al. argue that inasmuch as the "bare theory" exhibits feature (1), it is not really so "bare" after all and actually resembles what they term a "nonstandard" approach to interpreting quantum theory, that is, an approach in which something is added to the "bare theory" such as the stipulation that one observable is to be "preferred" over others, either in having an always-determinate value or at least in being a

---

[23] DeWitt (1970).
[24] Technically, this is described as Alice being in an eigenstate of "determinate measurement result," even if she is not in an eigenstate of one particular result or another.

"default" for determinacy. (Bohm's interpretation, to be discussed below, is an example of a nonstandard approach of this type, in that position is the privileged observable.) And regarding (2): as Bub et al. point out, other nonstandard approaches can give an account of how Alice could report not only that she had some definite belief about the result she observed, but what that result was. So, in their analysis, the bare theory falls short, both of actually being "bare" and of actually solving the measurement problem.

As for the DeWitt full-blown MWI version of the Everett approach, a major challenge is to explain what the quantum mechanical weights, or probabilities, mean if each outcome is actually *certain* to occur in some branch (world) or another. Doesn't the fact that something comes with a probability attached to it mean that there is some uncertainty about the actual outcome? The basic position of MWI – that all outcomes will certainly occur – has led to rather tortuous and esoteric arguments about the meaning of probability and uncertainty.[25]

But the situation may yet be worse for Everettian interpretations. Recently, Kent (2010) has argued that the whole program of deriving the Born Rule[26] from a decision-theoretic approach based on the presumed strategies of rational inhabitants of a "multiverse" (a MWI term for the entire collection of universes) is suspect. Any presumed strategy of a "rational" agent is no more than that – a probably sensible strategy among other possibly sensible strategies, and is therefore not unique. As Kent (2010) puts it:

The problem is that abandoning any claim of uniqueness also removes the purported connection between theoretical reasoning and empirical data, and this is disastrous for the program of attempting to interpret Everettian quantum theory via decision theory. If Wallace's arguments are read as suggesting no more than that one can consistently adopt the Born rule if one pleases, it remains a mystery as to how and why we arrived at the Born rule empirically. *(p. 10)*[27]

Besides the dependence on assumptions about what a rational agent would do, many approaches to deriving the Born Rule in the Everettian scheme depend on assumptions about mind–brain correspondences which are highly speculative as well as explicitly dualistic. As Kent (2010) observes:

the fact that we don't have a good theory of mind, even in classical physics, doesn't give us a free pass to conclude anything we please. That way lies scientific ruin: any physical

---

[25] Greaves (2004, pp. 426–27) proposes giving up the idea that the Born probabilities associated with the set of possible outcomes implies uncertainty about which outcome will happen. Meanwhile, Wallace (2006, pp. 672–73) proposes giving up the idea that being probabilistically uncertain of something pertains to the occurrence of some objective fact (outcome).

[26] The Born Rule is the prescription for calculating probabilities; see note 13.

[27] Kent refers to Wallace (2006).

theory is consistent with any observations if we can bridge any discrepancy by tacking on arbitrary assumptions about the link between mind states and physics. *(p. 21)*

Nevertheless, it would seem that Everettian arguments for the emergence of the Born Rule are crucially based on just such assumptions.

1.3.3 Bohm's Interpretation

In a nutshell, David Bohm (1917–92) proposed that the measurement problem can be solved by adding actual particles, possessing always-precise positions, to the wave function. To distinguish these postulated objects from the general term "particle" which is often used to refer to a generic quantum system, I will follow Brown and Wallace (2005) in terming these postulated Bohmian objects "corpuscles." The "equilibrium" distribution of these corpuscles is postulated to be given by the square of the wave function, in accordance with the Born Rule. The uncertainty and indeterminacy discussed earlier is still present in the Bohmian account. However, it is epistemic (rather than ontological) since they do possess definite positions but we cannot know what their positions were prior to detecting a particular measurement result. That is, the knowledge we can have of corpuscle positions at any time before a given measurement is limited to the distribution given by the square of the wave function of the system of interest (e.g., an electron in a hydrogen atom) (see Figure 1.5). The wave function then acts as a guiding or "pilot wave" for the corpuscle, as first suggested by Louis de Broglie (1923).[28] At the end of a measurement, the wave function will still have various "branches" (corresponding to different possible outcomes), but the corpuscle will occupy only one of them, and according to Bohm's formulation, this determines which result will be experienced. Thus the idea is that the Bohmian corpuscle acts as a kind of "agent of precipitation" which allows for the experience of one outcome out of the many possible ones. In terms of measurement, Bohm argues that the "corpuscular" aspect of the measuring apparatus, on interacting with the measured quantum system, ultimately enters one of the distinct guiding wave "channels" of the wave function of the entire system (apparatus plus quantum system) created through the process of measurement, and this process singles out that particular channel as the one which yields the actual result. (Brown and Wallace call this the "result assumption."[29])

---

[28] As far as I know, there is no physical account of how the "guiding wave," which lives in a $3N$-dimensional configuration space (where N is the number of corpuscles), guides the corpuscle – which is postulated to live in physical space. In the interest of a "level playing field" for competing interpretations, this lacuna should be kept in mind when considering criticisms of TI asserting that no specific "mechanism" is given for how a particular transaction is actualized.

[29] Brown and Wallace, in their careful analysis of Bohm's seminal 1952 papers, comment in passing that Bohm apparently did not intend to "surpass" quantum theory – to propose, in their words, a theory with "truly novel predictions" (Brown and Wallace, 2005, p. 521). This may be a reference to the fact that the Bohmian approach

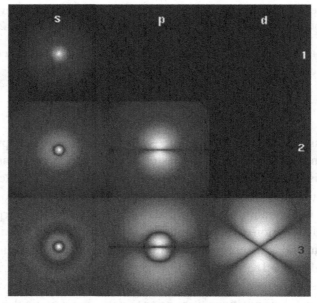

Figure 1.5 The squared wave function of an electron in various excited states of the hydrogen atom.
Source: https://commons.wikimedia.org/wiki/File:HAtom Orbitals.png.

1.3.4 Von Neumann's Projection Postulate

The formulation of John von Neumann, one of the pioneers of measurement theory in quantum mechanics, is not so much an interpretation as an analysis of the logical and statistical characteristics of the theory. It was von Neumann who first realized

amounts to a slightly different theory from standard quantum theory (cf. Valentini, 1992). The aspect of concern to me is the characterization of such a development as a "surpassing" of quantum theory and the implication that a good interpretation should make "novel predictions" (i.e., predictions that deviate from those of standard quantum theory). This language seems to imply that quantum theory is in need of improvement or remediation and that a proper interpretational approach should generate a "better" (different) theory. In contrast, I think nothing is wrong with the theory itself and that prevailing interpretational approaches have not gotten to the root of the measurement problem: namely, the need to include absorption as a real physical process generating advanced states (confirmations). Technically, this might be considered a different theory or at least a different formulation, but it is empirically equivalent at the level of probabilities to the standard theory. I do not believe that a successful *interpretation* (or formulation) needs to generate any novel predictions, but should provide a coherent and illuminating account of the theory itself, which effectively addresses the measurement problem. As a side note, an anonymous referee once commented in response to a statement like the preceding: "Since when has physics *not* dealt with difficult interpretational problems by changing the theory?" However, such changes were made not in response to *interpretational* problems, but rather to deal with the failure of a particular theory's *predictions*. For example, classical electrodynamics prior to relativity predicted that the speed of light should be dependent on the observer's motion. This prediction was refuted by the Michelson–Morley experiment. In contrast, the predictions of quantum theory are impeccable; it is probably the most strongly corroborated modern physical theory we have. What is at issue is arriving at a proper understanding of why the theory has the structure that it does, and to define measurement from within the theory. To modify the theory in an ad hoc way in order to get around these problems is, I believe, to fail to address the real scientific challenge it presents: What unexpected message does it convey about reality?

that the mathematical structure of the theory is a special kind of vector space (called a *Hilbert space*, in honor of the brilliant mathematician David Hilbert, who first defined it). While systems in classical mechanics can be represented mathematically as simple points labeled by their spatial position and momentum (technically, their coordinates in "phase space"), quantum systems have to be represented by rays in Hilbert space, which are objects that do not have simple coordinate-type labels and which reflect an infinitely expansive ambiguity as to the "actual" characteristics of the systems they represent. Roughly speaking, one can think of the classical phase space coordinatization as only one of an infinite number of ways to provide a coordinatization in Hilbert space.[30]

Von Neumann's view of measurement is often referred to as the "standard collapse approach," since it simply assumes that, on measurement, the state of the quantum system "collapses" (technically, it is "projected" onto a particular state corresponding to the type of measurement performed). He identified two different types of processes undergone by quantum systems: the "collapse" or "projection" that occurs on measurement he termed "Process 1," and the simple deterministic evolution of a system's state between measurements he termed "Process 2." Of course, he left unclear exactly what is supposed to precipitate the collapse of "Process 1," and this remains part of the measurement problem. (An additional problem traditionally associated with collapse is that it appears to be in conflict with relativity, since it seems to call for a preferred frame of simultaneity denied by relativity. On the other hand, TI's approach to collapse is harmonious with relativity, as will be demonstrated in Chapters 5 and 8.)

As I will discuss later in the book, the question of what triggers collapse cannot be properly answered unless absorption accompanied by non-unitarity is included in the dynamics. Without it, there is no clear "stopping point" at which a measurement can be regarded as completed (this was alluded to in Section 1.2.3), and all we have are vague "irreversibility" arguments that attribute *apparent* collapse to environmental dissipation or to "consciousness," but never really allow for a genuine physical collapse. At some point, an arbitrary "cut" is made at which the measurement is declared finished, "for all practical purposes" (a phrase which is often abbreviated "FAPP" in honor of John Bell, who introduced the term as an expression of derision[31]). This arbitrary demarcation between the microscopic systems clearly described by quantum theory and the macroscopic objects which

---

[30] This observation reinforces the point made in note 28: the mathematical structure of the theory is qualitatively different from that of classical mechanics, in a very striking way. To understand the physical reason for this mathematical structure, I suggest, is the real interpretational challenge. The Everettian approach is one way of embracing the challenge, but I think it fails because it disregards half the dynamics (the advanced solutions to the complex conjugate Schrödinger equation) and cannot provide a physical (as opposed to epistemic/ statistical) explanation for the Born Rule.

[31] Bell introduced this term in his essay, "Against measurement" (Bell, 1990).

"measure" them is often referred to as the "Heisenberg cut" in view of Heisenberg's discussion of the issue (cf. Bacciagaluppi and Crull, 2009).

Under TI, with absorption taken into account, collapse occurs much earlier in the measurement process than is usually assumed, so that we don't need to include macroscopic objects such as Geiger counters, cats, or observers in quantum superpositions. This issue is discussed in Chapters 3 and 4.

1.3.5 Bohr's Complementarity

Neils Bohr, one of the pioneers of quantum theory along with Werner Heisenberg, developed a philosophical view of the theory that he termed "complementarity." Complementarity has been the subject of enormous quantities of research and elaboration. Readers interested in a detailed critique of Bohr's formulation are invited to consult Kastner (2016b). Bohr's views will be described in more detail in Chapter 2. In brief, Bohr considered the properties of quantum systems to be fully dependent on what observers choose to measure, in that the experimental setup determines what sorts of properties a system can exhibit.[32] The Kantian flavor of his approach (after German philosopher Immanuel Kant) consists in denying that it is even meaningful to talk about the nature of the systems "in themselves," apart from their being observed in a macroscopic context. Based on Bohr's designation of such questions as "meaningless" or as beyond the domain of legitimate inquiry, his approach has been sardonically referred to as "shut up and calculate" (SUAC), a phrase coined by David Mermin (1989).

1.3.6 Ad hoc Nonlinear "Collapse" Approaches

So-called "spontaneous collapse" approaches such as that first proposed by Ghirardi, Rimini, and Weber (GRW) (Ghirardi et al., 1986) impose an explicit theoretical modification on the mathematics of the standard theory – an additional nonlinear term in the usual dynamics – in order to force a collapse into a determinate state. The added nonlinear component takes a poorly localized wave function and compresses it. This approach is explicitly and unapologetically ad hoc and faces several problems, among them the following. (1) A wave function that is compressed in terms of position must, by the uncertainty principle, gain a large uncertainty in momentum and therefore energy, which opens the door for observable effects, such as a system suddenly heating up – such effects are never observed. (2) Such collapses could occur only rarely; otherwise, the well-

---

[32] Bub has shown (1997) that complementarity can be viewed as a kind of "preferred observable," "no-collapse" approach, akin to the Bohmian interpretation which views position as the preferred observable. Bohr's preferred observable is whatever is measurable using the experimental setup.

corroborated normal evolution of the wave function would be noticeably disturbed. So it is not clear that their occurrence would be sufficient to account for the determinate results we see. Such "compression of the wave function" approaches are generally acknowledged as not viable, even by proponents of nonlinear collapse, and Tumulka (2006) has proposed a variant which purports to avoid some of the pitfalls known to afflict the original GRW approach.

Tumulka's proposal, a "relativistic flash ontology" version (rGRWf), avoids the compression problem (1) cited above. However, rGRWf still involves a physically unexplained and ad hoc "collapse" mechanism, and evades what I believe is the central interpretational issue of explaining why the theory has the mathematical Hilbert space structure that it does (see notes 28 and 29). In addition, in order to be reconcilable with relativity, rGRWf ultimately appeals to time symmetry. TI already makes use of time symmetry without needing to make any ad hoc change to the basic theory. I deal with this issue in more detail in Chapter 6.

1.3.7 Relational Block World Approaches

The term "block world" refers to a particular kind of ontology[33] in which it is assumed that spacetime itself exists as a "block" consisting of past, present, and future events. The block is unchanging and it is only our perception of it that seems to involve change as we "move" along our worldline. Such a view seems implied by relativity, and some researchers have proposed that quantum theory should be interpreted against such a backdrop. The challenge in doing so lies in explaining why the unitary evolution of a particular quantum state "collapses" to a particular result. Adherents of this view propose that such events simply correspond to a discontinuity of the relevant worldlines: that it is just a "brute fact" about nature that such discontinuities must exist.

This principle of a spacetime block with uncaused (primal) discontinuities was pioneered by Bohr, Mottelson, and Ulfbeck (BMU), who say (Bohr et al., 2003):

The principle, referred to as genuine fortuitousness, implies that the basic event, a click in a counter, comes without any cause and thus as a discontinuity in spacetime. From this principle, the formalism of quantum mechanics emerges with a radically new content, no longer dealing with things (atoms, particles, or fields) to be measured. Instead, quantum mechanics is recognized as the theory of distributions of uncaused clicks that form patterns laid down by spacetime symmetry. (abstract)

BMU take macroscopic "detector clicks" as primary uncaused events and refer to atoms as "phantasms." Thus they are explicitly antirealist about quantum objects.

---

[33] "Ontology" refers to what is assumed to exist, what is real.

BMU's approach has been developed more recently into a "relational block world" (RBW) interpretation by Silberstein, Stuckey, and Cifone (Silberstein et al., 2008). RBW advocates take spacetime relations and their governing symmetries as fundamental and attempt to derive a version of quantum mechanics based on this ontology.[34] One basis for criticism of RBW is that it makes fundamental use of dynamical concepts such as momentum while denying that those concepts refer to anything dynamical.[35]

1.3.8 Statistical/Epistemic Approaches

Some researchers (e.g., Spekkens, 2007) have been investigating an approach in which the quantum state reflects a particular preparation procedure but does not necessarily describe the physical nature of the quantum system under study. This implies that the quantum state characterizes only our knowledge; "epistemic," from the Greek word for "knowledge," is the technical term used. The statistical aspect consists in connecting a particular preparation procedure to a particular distribution of outcomes. The key feature distinguishing this "statistical" approach from the "hidden variables" approaches – such as Bohm's theory – is that in the former the quantum state is not uniquely determined by whatever "hidden" properties the quantum system possesses. In contrast, a quantum system under the Bohm theory is physically described by its wave function as well as an unknown position x of the postulated particle associated with the wave function; there is only one wave function that can be associated with these properties, even though the same wave function can be associated with another system with a different particle position x'.

However, a theorem by Pusey et al. (2011) casts serious doubt on epistemic/statistical approaches. It shows that, given some fairly weak assumptions, the statistics of a system whose state is not uniquely determined by its physical properties can violate the quantum mechanical statistical predictions.[36] The implication is that the quantum state really does describe a physical system, not just our knowledge of our preparation procedure.

---

[34] I do, however, share RBW's rejection of a "building block" ontology: the empirical world is a network of transactions, not collections of primitive individuals.

[35] For example, in RBW, experimental configurations are described by symmetry operators such as the translation operator $T(a) = \begin{pmatrix} e^{-ika} & 0 \\ 0 & e^{ika} \end{pmatrix}$, because momentum k is the generator of spatial translations. But, in RBW, there are no entities that possess momentum. It thus remains unclear what dynamical terms such as "momentum" refer to, in an adynamical account such as RBW.

[36] Granted, one of those assumptions is that there is no retrocausality. However, it is unclear to what extent adding retrocausality about an underlying ontology would help to support the basic statistical/epistemic program, which is to restore a more commonsense (i.e., classical) interpretation of quantum states than appears to be available from being realist about quantum states. If one is going to admit retrocausal influences anyway, then why not embrace a straightforward realist time-symmetric interpretation such as TI?

1.4 Quantum Theory Presents a Genuinely New Interpretational Challenge

Some researchers take the point of view that the appropriate response to quantum theory's apparently intractable puzzles is to adopt a strictly empiricist, pragmatic point of view, for example, to simply say that there is no physical explanation for the puzzling behavior of quantum objects as reflected in the theory, that nature simply "refuses to answer" the questions we try to pose about that behavior. One such approach, "Qbism" (proposed by Christopher Fuchs and David Mermin), holds that quantum theory is no more and no less than an instruction manual for predicting our experiences (a form of instrumentalism). Such approaches are variants of the Bohrian/Kantian view that people can gain knowledge *only* of the phenomenal level of appearance; that quantum theory might permit us to "knock at the door" of the subempirical, subphenomenal world but that the door must remain forever closed. This approach, I believe, is to evade a genuine, nontrivial interpretational challenge posed by the theory; that is, it admonishes us to renounce the idea that physical theories can describe nature itself.

While I certainly agree with the idea that quantum theory has an unexpected message, I think that message *is* one about reality – like all profoundly corroborated and powerfully predictive theories – and that the challenge is to figure out what the theory is telling us about reality. As this book will reveal, I think it is an exciting, strange, and indeed revolutionary message; certainly more interesting and revolutionary than the notion that theories of small things can only be about subjective knowledge or only about appearances. It was the behavior of hydrogen atoms that inspired Heisenberg to arrive at his first successful version of quantum theory. Clearly, the theory he arrived at was about those atoms and not just about his knowledge, since without reference to, and guidance from, those atoms he would never have constructed the theory. That is, the theory's structure was *driven by the behavior of atoms*. (Yes, the "observable behavior" of atoms, but the conclusion that the theory is only about our knowledge of them does not follow; this point will be explored further in the following chapter.) Jeeva Anandan underscored this point when he said:

[Quantum] theory is so rich and counterintuitive that it would not have been possible for us, mere mortals, to have dreamt it without the constant guidance provided by experiments. This is a constant reminder to us that nature is much richer than our imagination. (Anandan, 1997, p. 31)

The true puzzle of quantum theory is that there are physical entities beyond our power to perceive directly in the ordinary way and that they behave in strange and amazing ways. This is not just anthropocentrically about "our knowledge"; it is also about the physical entities. What are they saying to us? Heisenberg listened, and in the next chapter I will further explore his initial insights.

2

The Map versus the Territory

In this chapter, I consider some general issues of interpretive methodology, to present to the reader the motivation behind the new TI. I then argue in favor of a realist approach as opposed to an instrumental one.[1]

First, I should note that I offer an interpretive reformulation of what MacKinnon (2005) calls a "functioning," or informal theory: nonrelativistic quantum mechanics and its extension into the relativistic domain via quantum field theory. Since functioning theories are often inherently "untidy" (in either a mathematical or conceptual sense or both), philosophers of physics often engage in "rational reconstruction" of theories in order to render them more logically self-consistent in the hopes that the resulting formal theory will better lend itself to an unambiguous interpretation.[2] However, MacKinnon observes that in general, history does not support the notion that such recast, formalized theories lead to robust ontological insights. He instead characterizes the interpretive task as one of "find[ing] a way of relating philosophical questions about epistemology and ontology to functioning physical theories, rather than idealized constructions" (p. 4). That, in a nutshell, is the aim of the present work, although I believe that ultimately the model proposed herein is significantly more "tidy," ontologically self-consistent, and formally unified than conventional approaches to quantum theory.

---

[1] This chapter primarily critiques instrumentalist views; however, many so-called realist approaches to quantum theory contain unacknowledged instrumentalist or positivist-flavored assumptions about what the term "reality" means (such as "real" equals "empirically detectable"), so the discussion herein is relevant to those as well.

[2] As an example of this "untidiness," nonrelativistic QM and its relativistic extension might well be considered two different functioning theories, yet clearly they must describe the same reality and therefore presumably must be parts of a larger theory. A point of contact is found in Zee's observation (Zee, 2010, p. 19) that nonrelativistic quantum mechanics can be obtained in the Lagrangian formulation as a 0+1-dimensional quantum field theory.

2.1 The Irony of Quantum Theory

The original inception of quantum theory and the course of its subsequent evolution contain a deep irony. To appreciate this irony, we first need to revisit a bit of history.

2.1.1 Heisenberg's Breakthrough

A major breakthrough in quantum theory was achieved in 1925 through a decision by German physicist Werner Heisenberg to let go of certain preconceived metaphysical assumptions about the nature and behavior of matter: specifically, that we could picture electrons as little particles – corpuscles in the Greek (Democritan) conception – orbiting an atomic nucleus. Facing a theoretical impasse in accounting for atomic phenomena, he renounced these classical *anschaulich* (German for "picturable")[3] assumptions and retained only observable quantities such as energy differences and radiation frequencies, which could be measured and recorded as hard data. These he entered into arrays which he sardonically termed "laundry lists," and which his then-teacher Max Born would soon realize were matrices (arrays of numbers in a form well known in mathematics). Thus was born Heisenberg's "matrix mechanics" version of the theory, which successfully predicted the experimental (spectral) data arising from observations of the hydrogen atom. Subsequent development would eventually lead to a powerful, empirically successful theory which could be expressed in different forms (probably the best known being the Schrödinger wave mechanics, based on Erwin Schrödinger's celebrated equation), and whose formal structure was described, as von Neumann had first noticed, by an abstract mathematical space called Hilbert space.

2.1.2 Bohr's Antirealism

However (as observed in Chapter 1), nearly a century later, researchers are still deeply puzzled about how to interpret the theory, in the sense of understanding what it says about reality (if anything). Most physicists and philosophers of physics are aware that Heisenberg's breakthrough came as a result of renouncing his preconceived metaphysical assumptions, and many of them (including, most notably, Heisenberg's fellow quantum theory founder Niels Bohr) have taken from this fact what I believe is the wrong lesson: they have renounced realism with regard

---

[3] The term *anschaulich* presupposes that "picturable" means the usual classical picture of corpuscles following determinate trajectories. This assumption is contested in the present account: physical processes could be "picturable" in terms of an entirely different kind of picture.

to quantum theory. That is, the idea that there was some understandable, underlying physical reality described by quantum theory tended to be viewed suspiciously, as a misguided impulse to drag in metaphysical baggage that Heisenberg's approach had discredited as inappropriate methodology. Probably nobody says this more emphatically than Neils Bohr: "There is no quantum world. There is only an abstract physical description. It is wrong to think that the task of physics is to find out how nature is. Physics concerns what we can say about nature."[4]

The above sentence by Bohr presupposes that nature can only be talked about using classical concepts, that is, the very "picturable" notions that Heisenberg had renounced in order to arrive at his matrix formulation of quantum theory. In effect, Bohr presumed that physics cannot "say how Nature is," even though quantum theory, as a formal structure, *may be doing just that*, albeit not in the traditionally picturable manner. Bohr's positivistic prohibition on "finding out how nature is" was not necessarily heeded by everyone, but it had, at the very least, a chilling effect on interpretive inquiry.[5]

In particular, Bohr's legacy is alive and well among many practicing physicists, whose job it is to calculate experimental predictions and analyze results, and who tend to regard efforts by philosophers of physics to "find out how nature is" to be a misguided waste of time. Many of them approach interpretational puzzles of quantum theory from the kind of deflationary, "debunking" view alluded to at the end of the previous chapter. Of course, nobody is to be faulted for choosing not to be realist about physical theory, especially when it is not in their job description to do so. But the main thesis of this work is that, contra Bohr, it is perfectly *reasonable* to be realist about the subject matter of quantum theory, and that it is perfectly possible to "find out how nature is," as long as we don't expect it to be "classically *anschaulich*" and are willing to entertain some new and unexpected ideas of how nature might be (analogous to the strange specter of energy having to be "quantized," which led to Max Planck's successful derivation of the blackbody radiation spectrum).[6]

2.1.3 Einstein's Realism and a Further Irony

Einstein, as is well known, completely disagreed with Bohr's approach. His motivation was, in his own words, to "know God's thoughts."[7] Yet, ironically, a

---

[4] As quoted in Petersen (1963).

[5] Some of Bohr's most famous pronouncements about the meaning and implications of quantum theory depend heavily on optional metaphysical and/or epistemological claims treated by him as obligatory, or are simply self-contradictory (for some examples, see Kastner, 2016b).

[6] Planck had introduced a discrete sum of finite energy chunks as a calculational device only. When he tried to take the limit of the sum as the size of the chunks approached zero, he got back the old – wrong – expression. The chunks had to be of finite size in order to get the correct prediction.

[7] "I want to know God's thoughts. The rest is details." Widely attributed to Einstein.

similar antirealist tendency has recently arisen based on the methodology Einstein used in formulating his theory of special relativity. Einstein famously arrived at his theory by thinking in terms of what someone could actually measure with (idealized) rigid rods and clocks, and concluded that one needed to renounce certain metaphysical notions about space and time: in particular, Newton's view that space and time are absolute, immutable "containers" for events. What is less often remembered is that Einstein also used formal theoretical assumptions: in particular, he demanded the invariance of electromagnetism, requiring that the theory not be dependent on an observer's state of motion. But the prevailing message of relativity came to be that there is no such thing as absolute simultaneity or absolute lengths of objects and that these concepts were metaphysical ballast to be jettisoned. Einstein's renunciation of such absolute metaphysical concepts is often amplified, like Heisenberg's renunciation of the trajectory concept, into a universal doctrine that any notion of an underlying (i.e., subempirical) reality is to be eschewed.

However, not only is this an inappropriate lesson to take from these theoretical achievements; it is not even consistently applied: most researchers (and especially physicists) continue to be thoroughgoing realists about spacetime, viewing it as a fundamental substantive "container" or backdrop which not only underlies all possible theoretical models but which even has causal powers to "steer" particles on trajectories.[8] And one must note the additional irony that the notion of "trajectory" persists, despite the widespread view that fundamental reality should not be considered "picturable."[9]

2.1.4 Theory Construction versus Theory Interpretation

The point generally overlooked in the trend described above is that theory formulation/discovery is an entirely different process from that of theory interpretation. We need to distinguish between (1) the valid point that preconceived metaphysical assumptions can serve as a barrier to theory *invention or discovery*, especially when a successful new theory cannot be based on such assumptions, and (2) realist *interpretation* of an existing empirically successful theory as a way of discovering *new* features of reality uncovered by that theory.

---

[8] The commonplace notion that spacetime has causal power to steer particles is subject to sustained and cogent criticism by Harvey Brown (2002).

[9] For example, many discussions of the "two-slit" experiment and similar experiments, in which the state of a single quantum is placed into a superposition by a half-silvered mirror or other means, are centered around so-called which-way information. This term is heavily laden with the presumption of a determinate trajectory: surely, if one talks about "which-way information," one tacitly assumes that the entity under study went either one (spacetime) way or the other, that is, pursued a trajectory. So, even though perhaps not always intended, its use smuggles in a supposedly renounced classical metaphysical picture.

The deep irony of quantum theory, I suggest, is that its discovery was made possible by the renunciation of a then-realist approach and attendant metaphysical baggage; yet when interpretationally queried from a realist perspective in the proper way, quantum theory can open the way to an entirely new and richer understanding of physical reality: a strange new kind of model that we could not have discovered without first letting go of inappropriately classical metaphysical concepts. In making this claim, I invite the reader to reflect on the insightful quote at the end of Chapter 1, by the late Jeeva Anandan.

2.2 "Constructive" versus "Principle" Theories

What do I mean by querying a theory "in the proper way"? In order to address this, I first need to review an important distinction in theory type: "constructive" versus "principle" theories. Simply put, a constructive theory is one based on a model. A famous example is the kinetic theory of gases, which represents the behavior of gases in terms of small, impenetrable spheres in collision with one another and the walls of their container. By applying known physical laws to this model, James Clerk Maxwell and Ludwig Boltzmann were able to deduce the large-scale thermodynamic behavior of gases; for example, Boyle's Law relating temperature, pressure, and volume ($PV = nRT$).[10] Such a "constructive" theory is powerful and illuminating because it allows us to understand the "nuts and bolts" of what is really going on at a level beyond ordinary experience, that is, beneath the phenomenal level of appearance. That is what Einstein meant when he talked about wanting to "know God's thoughts." He didn't just want to know about how God's creation *appears* and to be able to analyze, classify, and predict those appearances; he wanted to know how it all works beneath the merely phenomenal level, "to boldly go" where Bohr summarily pronounced that nobody should be able, nor wish, to go.[11]

In contrast, a "principle" approach to theory development lacks a physical model. It starts from an abstract principle or principles that serve to constrain the form that the theory can take, and then fits the theory, with the help of mathematical consistency and basic physical laws such as energy conservation, to empirical observation. Relativity was a principle theory, and Einstein was very dissatisfied with this aspect of it. He felt that only a constructive theory, with its attendant illuminating model, provided genuine insight into "how nature really is." Similarly, quantum mechanics was a principle theory, as we can see by the fact that Heisenberg had to explicitly jettison the models he was trying to work with

---

[10] A comprehensive and very readable account of this scientific episode is found in Brush (1976).
[11] With apologies to Gene Roddenberry.

(i.e., his erroneous metaphysical pictures of how atoms behaved), and to work only with empirical observations that served to constrain the form of the theory. Before that, Planck used a purely mathematical trick – summing over discrete energy levels instead of assuming energy was a continuously variable quantity – to obtain the correct empirical result for blackbody radiation (see, e.g., Eisberg and Resnick, 1974, section 1.1 and especially p. 14 for a clear account of how this phenomenon presented a fatal problem for classical electromagnetism and forced the invention of quantum theory). His desperate resort to this tactic led to the discovery of Planck's constant, h, the fundamental physical constant which characterizes the smallest unit of action (units of energy times time or momentum times length). Thus his approach to the discovery of the new theory was principle-based (i.e., using formal mathematical considerations), not model-based.

2.3 Bohr's Kantian Orthodoxy

Now, as noted above, Bohr was perfectly content with the idea that quantum mechanics was a principle theory. He assumed from the way that the theory was arrived at – by rejecting a model that didn't work – that there *can be no model* for quantum theory, that is, no way of picturing "how nature is." In other words, Bohr elevated the fact that one cannot apply *classical* model-making to a *nonclassical* realm into a broad-brush policy that, at the quantum level, one should not try to find models of *any* kind. He assumed that if one cannot have a classical model, there can be no model, and that quantum theory represents the end of the scientific search for understanding of the physical world in a realist sense: that is, independently of how we happen to be looking at it.

Put differently, he assumed that classical modeling is *equivalent* to giving a realist account of micro-reality and that one therefore cannot give such an account. This formulation is rebutted eloquently by Ernan McMullin, who wrote:

[I]maginability must not be made the test for ontology. The realist claim is that the scientist is discovering the structures of the world; it is not required in addition that these structures be imaginable in the categories of the macroworld. (as quoted in Ladyman, 2009)

Bohr's formulation depended heavily on appealing to phenomenal and epistemic notions, such as the fact that in order for scientists to communicate their results, they had to be able to talk about pointer readings, thereby working with classical phenomena and speaking in "classical language." This is true, of course – it is the means by which all physical theories are tested and corroborated. However, it does not follow from limitations on scientists' interactive verbal language requirements that the mathematical structure of the theory has no real, objective referent. Bohr simply jumped to an unwarranted conclusion in this regard, based on his tacit assumption that any reality describable by a physical theory must be classical in nature.[12] As observed by McMullin, the interpretational question is whether the theory's formal content (e.g., the Schrödinger equation and its solutions) has some physical referent, regardless of whether or not that physical referent can be directly observed (or described in macroscopic, classical terms).

At this point it is useful to acknowledge a distinct similarity between Bohr's thought and the work of German philosopher Immanuel Kant. Kant proposed that reality has two fundamental aspects: (1) the world of appearance and (2) the "thing-in-itself" (or "noumenon"), which he held was unknowable.[13] For an accessible introduction to the problem of gaining knowledge of the "thing-in-itself," the reader is encouraged to consult chapter 1 of Bertrand Russell's *The Problems of Philosophy*, in which the author considers an ordinary table and presents a convincing case that the table itself, apart from any perception of it, is a deeply mysterious object, "if it exists at all." (For an updated version of this epistemological puzzle, see Section 7.5.) Kant also proposed that there are "categories of experience" that make knowledge of the world of appearance possible, and which are the only means through which knowledge is constructed.[14] Knowledge, for Kant, was *only* about item (1), the world of appearance; recall that part of the definition of (2), the thing-in-itself or underlying reality, was that it was intrinsically unknowable. Among the "categories of experience" were concepts like space, time, and causality. In particular, Kant proclaimed that Euclidean space was an a priori category of understanding, meaning a necessary concept behind any knowable phenomenon – an assertion which, it should be noted, has since been decisively falsified by relativity's non-Euclidean accounts of spacetime.

Bohr seems to have assumed, much like Kant, that all knowledge obtained by way of physical theories applies only to the world of appearance and that the "classical modes of description" are required for all knowledge. So Bohr's "classical modes of description" play the same role as Kant's "categories of experience." Bohr, in essence, proclaimed that while quantum theory might have placed us just at the doorstep of the "noumenal" realm, the nature of the theory required that we could not gain knowledge about it and that, moreover (as a "normative" principle echoed by modern day "Qbists"), it would be scientifically and methodologically unsound to think that we should try to do so, as reflected in his previous quote. By "abstract quantum mechanical description," Bohr

---

[12] For a detailed critique of Bohr's unnecessary jump to instrumentalism about quantum theory, see Kastner (2017a).

[13] Kant often used the "thing-in-itself" interchangeably with the term "noumenon," a Greek term which translates roughly as "object of the mind." Kant's division is very similar in structure to Plato's division, as the reader will recall from Chapter 1.

[14] Kant's ideas discussed here were presented in his *Critique of Pure Reason* (1996).

preemptively denied that the formalism could be referring to anything physically real, thus effectively relegating it to a linguistic or computational device. As noted above, this assumption can and should be questioned.

It has often been pointed out (e.g., by Bohr and Heisenberg) that in general there can be no mechanistic, deterministic account of individual microscopic events. This fact is often referred to in terms of "quantum jumps" that cannot be predicted, even in principle. Yet a realist understanding of micro-reality *need not* take the form of a detailed mechanical account of an individual event – the entity that remains elusive to causal description, as Anton Zeilinger notes.[15] To assume, like Bohr, that a realist understanding must be in terms of the usual "classical," causal account is to unnecessarily limit ourselves to a pseudo-Kantian "category of experience." Many of these have already been shown to be obsolete by scientific advance, as noted above. The new realist understanding may not be in terms of causal, mechanistic processes. It may instead encompass a fundamental indeterminism at the heart of nature, but one which is well defined in terms of the conditions under which it occurs – in contrast to prevailing orthodox interpretations which suffer from an ill-defined micro/macro "cut" (as discussed in Section 1.3.4). The new understanding offered here is a rational account, in the sense of being well-defined and self-consistent, even while it lacks certain features, such as determinism and mechanism, that have been traditionally assumed to be requirements for an acceptable scientific account of phenomena.

Thus, as alluded to above, Bohr's famous conclusion that "It is wrong to think that the task of physics is to find out how nature is" is a logical fallacy. It simply does not follow logically that the failure of classical model-making entails that no model of any kind is possible.[16] As noted above, one may regard Bohr as making the same kind of mistake as Kant when the latter presumed that there can be no knowledge of a realm that is not based on a Euclidean space. While it may be true, as a matter of contingent fact, that there is no adequate model, there is no reason that a failure of a particular sort of inappropriate model should be turned into a general prohibition against modeling. On the contrary, a principle theory can provide truly groundbreaking insights into new aspects of reality: it can ultimately lead us to a *new* kind of model, one so utterly different from how we are used to thinking about reality that we could not have approached it directly, "from the ground up" so to speak, but had to arrive at it through an indirect route, "top down," as Heisenberg did. This is the insight expressed by Jeeva Anandan, quoted at the end of Chapter 1.

---

[15] Zeilinger (2005). However, Zeilinger's definition of "realism" is what I would call "actualism"; see Chapter 8 and Kastner (2019c, chapter 5).

[16] I recognize that Bohr adduced Kantian epistemological reasons for his prohibition against modeling in quantum theory, but I reject those as well. Specifically, it will be argued later in this chapter that the promise of quantum theory is to give us a glimpse of the "noumenal" realm, so I will be rejecting the Kantian claim that all knowledge must be restricted to phenomena.

2.4 The Proper Way to Interpret a "Principle" Theory

So, what is the "proper way" to interpret such a principle theory, one that was developed without reference to any model? To answer this question, let's turn to a famous dictum by Bryce DeWitt, who presented it as the essential motivation for his development of the Everett interpretation into what became known as the many worlds interpretation:[17] "The mathematical formalism of the quantum theory is capable of yielding its own interpretation" (DeWitt, 1970).

I take this to mean that the formalism resulting from whatever methodology was needed to develop an empirically successful theory – especially a principle theory like quantum mechanics, which was not based on prior construction of a model – has features that may well point to heretofore hidden or unnoticed features of reality. A perfect case in point, again, is Planck's stumbling upon quantized energy because his empirically successful quantitative theory said so, not because he wanted it that way. Since the features of an empirically successful principle theory are (apparently) not something we could have thought of unaided, they are not available to us as a possible model, and we (like Heisenberg) have to proceed without their help, "groping in the dark," so to speak, aided only by previously established physical principles, mathematical consistency, and empirical data to guide us to the form of the theory.

Heisenberg, in choosing to "listen to reality" by renouncing his previous unhelpful metaphysical assumptions, wrote down the "laundry list" formalism (matrix mechanics) that turned out to be a useful instrument for predicting observations arising from the microscopic systems he was studying. But, as argued above, it does not logically follow that all there is to reality is those abstract "laundry lists." A possibly useful analogy here is a map to some buried treasure: Heisenberg, through his choice to adopt a Zen-like "beginner's mind" approach to the phenomena under study, stopped listening to his own ineffective ideas and began to listen to the message of reality instead, as encoded in the phenomena. Thus, he was able to "hear" what reality was trying to tell him by writing down what became a useful "map." The realist impulse that underlies and motivates all fundamental scientific advance is to acknowledge that there is some reason, however obscure, that such a theoretical "map" allows us to predict phenomena. Unless we wish to believe in miracles or coincidence as the explanation for the success of a theory like quantum mechanics, or to deny that theory success even needs explaining, which is to retreat from the deepest aspects of the scientific and philosophical mission, we are obligated to acknowledge that the "map" reflects something about reality – however utterly new and unfamiliar.

---

[17] As mentioned in Chapter 1, it is my view that MWI advocates overlook part of the formalism (advanced solutions).

Another analogy for the inspiration leading to a successful principle theory is in the realm of psychology and interpersonal relationships. Successful mediators know that conflicts can be resolved when the parties are helped to let go of their own preconceived notions, desires, or requirements for the other person and start to listen to what the other person is saying. More broadly, a socially effective person has the ability to be receptive to the messages from their environment and the flexibility to adapt to the meaning of the messages, that is, to let go of preconceived notions about "how things should be" and to behave in ways that are more appropriate and fruitful. But they don't conclude from that that there is nothing further to be learned about that other person or situation, or that there is nothing beyond those messages they heard which allowed them to behave more effectively. A new way of behaving is more "fruitful" because there is *something there yielding fruit*. Heisenberg's approach exemplifies, albeit in a different context, the behavior of a successful person in social relationships. He stopped presuming and started listening, and was able to write down a very useful "map." We should not mistakenly conclude from his methodological success that there is no more to reality than that map.

So, as will be developed in later chapters, the proper way to interpret the theory is to "listen" carefully to its unexpected mathematical features. A crucial step was made by Max Born, who linked the absolute square of the Schrödinger wave function[18] to something empirical, if only statistical: this quantity could be seen to function as the probability of observing the associated property when one conducted a measurement of the system. His finding became known as the "Born Rule," and it is the fundamental empirical link between quantum theory and the world of phenomena. As noted in the previous chapter, in most prevailing interpretations, the Born Rule either is simply assumed as part of the mathematical machinery that does not merit or require explicit interpretation or is given a pragmatic, "for all practical purposes" (FAPP) account which, in my view, fails to do it justice as the crucial link between theory and concrete experience. The Born Rule constitutes a deep mystery for all prevailing interpretations; there would appear to be no straightforward ontological (i.e., non-epistemic, nonstatistical) explanation for it in any interpretation other than TI.[19]

---

[18] More generally, the probability is the square of the projection of the quantum state onto a particular classically observable property, for example, position or momentum.

[19] As noted in Chapter 1, Bohmians claim that the Born Rule is obtained as the statistical distribution of particle positions. But this is only for the so-called equilibrium state of the subquantum level (i.e., the level of determinate positions). Since the Bohmian theory allows for the particle position distribution to deviate from the Born Rule, it is a different theory from quantum mechanics. Even if one viewed the "non-equilibrium" state as improbable or even impossible, the account is only statistical, which I view as a weaker kind of physical explanation. A further challenge for the Bohmian account is that particles are continually created and destroyed in the relativistic regime, which would seem to increase the likelihood of distributions that might deviate from

2.5 Heisenberg's Hint: A New Metaphysical Category

Heisenberg took a further step in "listening" to quantum theory when he made the following statement: "Atoms and the elementary particles themselves are not real; they form a world of potentialities or possibilities rather than things of the facts."[20] This assertion was based on the fact that quantum systems such as atoms are generally described by quantum states with a list of possible outcomes, and yet only one of those can be realized upon measurement. I think that he was on to something here, except that I would adjust his characterization of quantum systems as follows: they are real, but not *actual*. In his terms, they are something not quite actual; they are "potentialities" or "possibilities." Thus my proposal is that quantum mechanics instructs us that we need a new metaphysical category: something more concrete than the merely abstract (or mental), but less concrete than, in Heisenberg's terms, "facts" or observable phenomena.[21] The list of possible outcomes in the theory is just that: a list of possible ways that things could be, where only one actually becomes a "fact." This proposal is directly analogous to Planck's proposal, in view of the inescapable formal features of his theory, that energy is quantized.

The distinction between a quantum possibility and a fact is clarified in a comment that Heisenberg made later in his life (and will be further clarified in Chapter 7):

The probability wave of Bohr, Kramers, Slater ... was a quantitative version of the old concept of "potentia" in Aristotelian philosophy. It introduced something standing in the middle between the idea of an event and the actual event, a strange kind of physical reality just in the middle between possibility and reality. (Heisenberg, 2007, p. 15)

So, Heisenberg had arrived at a new kind of metaphysical understanding, a "picture," if you will, of the reality described by quantum theory. However, in view of his ambivalence about it – he was a practicing physicist, after all, and expected models to be based on "things of the facts"– he did not pursue this insight as a viable description of the underlying reality described by quantum theory. Among my goals in this work is to essentially pick up where he left off (this ontological exploration begins in Chapter 4).

A further important aspect of "listening to the formalism" of quantum theory is to acknowledge its time-symmetric (or at least "advanced") aspects. Specifically, it cannot be overemphasized – since the fact is habitually neglected – that *advanced*

---

the "equilibrium" configuration needed for its empirical equivalence to standard QM. Everettians give an epistemologically based account of the Born Rule which must refer to the knowledge of an observer.

[20] Heisenberg (1958, p. 186).

[21] This proposal is explored in some detail in Kastner et al. (2018).

(time-reversed) states necessarily enter into any calculation needed to obtain empirical content (i.e., probabilities for outcomes of measurements, or expectation values for the values of measured observables). Indeed, this overlooked fact is so important that I will elevate it to an interpretational maxim for any realist interpretation:

> *Maxim: Mathematical operations of a theory which are necessary to obtain correspondence of the theory with observation merit a specific (exact) ontological interpretation.*

This proposed maxim no doubt requires some elucidation. For one thing, TI's rival "purist" interpretation (i.e., the collection of approaches constituting the so-called many worlds interpretation based on Hugh Everett's proposal of 1957) does not adhere to it. As alluded to earlier, MWI addresses the Born Rule by epistemological or statistically approximate methods: by arguing, via decision theory, that a rational observer would choose to bet on outcomes obeying the Born Rule; by arguing that Everettian worlds violating the Born Rule have approximately zero measure; and so on. Similarly, the Bohm theory proposes that the distributions of Bohmian particles closely approximate that specified by the Born Rule. Now, in the absence of any mathematical property of the basic theory that could provide an unambiguous ontological basis for the Born Rule, such approximate and/or ad hoc approaches might be justified. But the theory *does* possess a specific mathematical object that can provide an exact ontological basis for the Born Rule: the set of advanced solutions which, under TI, are confirmation waves arising from the ubiquitous absorption processes neglected in other interpretations. Since absorption processes are physically present whenever there is a detection (the latter being a requirement for an observation), the advanced solution is the obvious mathematical entity to interpret as a component of the ontological basis for the Born Rule.

Despite the counterintuitive aspects of advanced states, I believe that truly hearing what the formalism is saying means taking seriously the idea that it describes something with advanced (as opposed to the usual retarded) qualities. This is where, in my view, TI improves upon Everettian interpretations which try to approach the formalism from a receptive, "purist" point of view, but which fail to notice that the advanced states are a crucial part of the theory with physical content that should not be neglected.

The transactional conceptual picture represents a parallel to that of Einstein's conceptual unification of the instrumental and pragmatic prerelativistic quasi-theories, as described by Zeilinger (1996):

It so happened that almost all relativistic equations which appear in Einstein's publication of 1905 were known already before . . ., mainly through Lorentz, Fitzgerald and Poincaré – simply as an attempt to interpret experimental data quantitatively. But only Einstein created the conceptual foundations, from which, together with the constancy of the velocity of light, the equations of the theory of relativity arise. He did this by introducing the principle of relativity, which asserts that the laws of physics must be the same in all inertial systems. I maintain that it is this very fact of the existence of such a fundamental principle on which the theory is built which is the reason for the observation that we do not see a multitude of interpretations of the theory of relativity. *(p. 2)*

The Born Rule equating the probability of a particular result to the square of the wave function is one of the equations allowing quantitative interpretation of experimental data in quantum theory, just as the Lorentz contraction allowed quantitative empirical correspondence in prerelativistic theories. The current multitude of competing "mainstream" interpretations of quantum theory (among these the Bohmian theory, ad hoc "spontaneous collapse," approaches, MWI) are all different ways of providing approximate, pragmatic, after-the-fact justifications for the Born Rule and the conditions of its application – showing that its use is consistent with the rest of the theory in some limit – rather than an *explanation for how it arises* naturally from the theory. In contrast, the conceptual picture of a transactional process is what allows the operational equation of the Born Rule to arise from the theoretical formalism, just as Einstein's postulates allow the Lorentz contraction to emerge as a natural consequence.

2.6 Ernst Mach: Visionary/Reactionary

I digress slightly here to discuss Ernst Mach, a prominent figure in nineteenth-century physics, because he probably exemplifies more than anyone else the irony discussed in this chapter. He exemplified, on the one hand, the virtue of humble submission and obedience to nature's empirical messages and, on the other hand, the philosophical mistake of assuming that those empirical phenomena are all there is or that knowledge cannot, or should not, go beyond them. As a strict empiricist, Mach insisted that all knowledge is based on sensation or observation – a position that of course confines any empiricist to knowledge about the world of appearance only. Yet it does not follow that *the only thing that exists* is appearances, as noted earlier. Here I endorse von Weizsaecker's dictum that "What is observed certainly exists; about what is not observed we are still free to make suitable assumptions. We use that freedom to avoid paradoxes."[22] (Descartes has more pungent remarks for the strict empiricist, as we will see shortly.)

Thus, while I agree with Mach's eliminativist[23] account of spacetime as fundamentally based on comparisons (i.e., I adopt a relational view of spacetime),

---

[22] Private communication, first quoted in Cramer (1986).
[23] A term meaning that the concept under study does not correspond to an independently existing entity or substance.

that does not mean that the interpretation of all physical theories which were discovered through the application of mathematical analysis to observations must be limited to subjective sensations, as Mach unnecessarily (and I believe mistakenly) concludes in the last clause below:

[W]e do not measure mere space; we require a material standard of measurement, and with this the whole system of manifold sensations is brought back again. It is only intuitional sense-presentations that can lead to the formulation of the equations of physics, and *it is precisely in such presentations that the interpretation of these equations consists. (Mach, 1914; emphasis added)*

Thus, Mach's justified insistence that theory *construction* be grounded in observation slides unjustifiably into categorical antirealism about possible unobservable entities pointed to by those theories. As noted previously, this is a logical and methodological error, unambiguously revealed as such when Mach's refusal to entertain the existence of atoms – because they were unobservable – was shown to have been on the wrong side of scientific progress. One can acknowledge that perhaps what we think of as "spacetime" can be understood in terms of the ordering of sensations (also known as material objects), but it does not logically follow that there is *nothing more to reality* than those sensations. The ordering we discover can be seen as an objective property of reality insofar as all our observations conform to it and it cannot be altered by purely subjective means (i.e., by imagining or desiring it to be different). Thus, objective reality may be something real, even if not directly observable, which is capable of giving rise to sensations (i.e., observations or actualized events). The unjustified assumption that because our knowledge of reality is derived largely from sensation, our interpretation of theories and our understanding of reality must be *limited* to accounts of sensation, is subjected to rather harsh criticism by Descartes in his *Treatise on Light*. I quote generously here, as Descartes takes a while to establish his point:

[T]he spaces where we sense nothing are filled with the same matter, and contain at least as much of that matter, as those occupied by the bodies that we sense. Thus, for example, when a vessel is full of gold or lead, it nonetheless contains no more matter than when we think it is empty. This may well seem strange to many *whose [powers of] reasoning do not extend beyond their fingertips and who think there is nothing in the world except what they touch.* But when you have considered for a bit what makes us sense a body or not sense it, I am sure you will find nothing incredible in the above. For you will know clearly that, far from all the things around us being sensible, it is on the contrary those that are there most of the time that can be sensed the least, and those that are always there that can never be sensed at all.

  The heat of our heart is quite great, but we do not feel it because it is always there. The weight of our body is not small, but it does not discomfort us. We do not even feel the weight of our clothes because we are accustomed to wearing them. The reason for this is clear enough; for it is certain that we cannot sense any body unless *it is the cause of some change* in our sensory organs, i.e. unless it moves in some way the small parts of the matter of which those organs are composed. The objects that are not always present can well do this, provided only that they have force enough; for, if they corrupt something there while they act, that can be repaired afterward by nature, when they are no longer acting. But if those that continually touch us ever had the power to produce any change in our senses, and to move any parts of their matter, in order to move them they had perforce to separate them entirely from the others at the beginning of our life, and thus they can have left there only those that completely resist their action and by means of which they cannot be sensed in any way. Whence you see that it is no wonder that there are many spaces about us in which we sense no body, even though they contain bodies no less than those in which we sense them the most. *(Descartes, 1664, chapter 4; emphasis added)*

Thus (in admittedly uncharitable language), Descartes argues that it is a mistake to assume that nothing exists beyond what we sense, as our material senses can only detect *change*, not entities that are always present or that are incapable of activating our sense organs. It is widely supposed that Descartes' metaphysics, which postulated a dynamic plenum rather than a void underlying observable matter, was a quaint piece of "moribund metaphysics" (to use Van Fraassen's term)[24] that was largely discredited by Newton's theories. Yet, arguably, Descartes can now be seen as having presaged the development of relativistic quantum theory, which has taught us that what Newton thought of as the "void" is far from empty.[25] So we would do well to reacquaint ourselves with Descartes' views on scientific methodology. We should also consider the von Weizsaecker quote above that "what is observed certainly exists; about what is not observed we are still free to make suitable assumptions." Such a "suitable assumption," as remarked earlier, was the existence of atoms. So despite Mach's insights into the importance of recognizing how our knowledge is obtained largely through sensation, he refused to countenance a crucial theoretical construct – the atom – that crucially led to major scientific breakthroughs. The lesson, I suggest, is to acknowledge that we should not let metaphysical preconceptions get in the way of observations and theory construction based on those observations, but we should *not* uncritically assume from the success of that approach that, as Descartes says, "there is nothing in the world except what [we] touch."

2.7 Quantum Theory and the Noumenal Realm

So what can be gained by exploring the possibility that certain aspects of the quantum formalism typically thought to have only operational significance

---

[24] For example, Van Fraassen (2004, p. 3).

[25] For example, there is continual particle/antiparticle creation arising from the vacuum. Overall, an astonishing amount of activity goes on in so-called empty space.

(e.g., dual states or bracs, denoted as $\langle \Psi |$) may indeed have ontological significance? Recall Zeilinger's observation that the "individual event" remains resistant to causal description, along with similar observations by the founders of quantum theory. For example, according to Jammer (1993), Bohr referred to such events, such as the inherently unpredictable transitions of electrons in atoms from one stationary state to another,[26] as "transcending the frame of space and time."[27] As discussed earlier in this chapter, Bohr regarded spacetime concepts (indeed, all "classical" concepts) as prerequisites for the endeavor of gaining physical knowledge of the world; thus, he explicitly restricted what counted as legitimate knowledge to that of the world of appearance, in Kantian terms. Yet the significance of his quoted remark is that it clearly implies *there are real physical events which transcend the boundaries of the observable universe*. For surely Bohr has to acknowledge that stationary states were instantiated in nature and that transitions between them did occur, as this much is empirically corroborated.

Recall that Bohr insisted that physics concerns "what we can say about nature." But what is the "we" in this context? Is it ordinary language? Or is it the mathematical language of our best theories? If the former, obviously it is very difficult, if not impossible, to talk about events which "transcend the frame of space and time." But even Bohr implicitly admitted, as noted above, that such events occur. Indeed, the very theory he helped invent is what led him to make this observation. Does that not, then, mean that the formal aspects of physical theory *can* point to heretofore unknown aspects of physical reality, however difficult it might be to talk about them – that physics *can* be more than what we can "say about Nature" in ordinary, classically *anschaulich* terms?

I believe that the answer to that question is "yes." The fact that quantum theory, in Bohr's words, seems to point to entities and/or processes "transcending the frame of space and time" means that quantum theory can reasonably be thought of as (at least in part) a theory about the noumenal realm.[28] That is, since concepts like space and time are considered vital for gaining and communicating knowledge about the world of appearance, processes that transcend those concepts must be processes belonging to the noumenal realm, which transcends the world of appearance. Therefore, I claim that the truly revolutionary message of quantum theory is *not* that we should stop asking questions about the nature of reality; on the contrary, the message is that quantum theory is offering a new and strange kind of answer about an aspect of reality traditionally pronounced "off limits" by Kant and those (like Bohr) who subscribe to the notion that physical theory can only be

---

[26] Stationary states are states whose wave functions do not change with time. An atom's discrete energy levels correspond to such states.

[27] As quoted in Jammer (1993, p. 189).

[28] More precisely, that the domain of quantum theory includes the noumenal realm as a component.

about the world of appearance. That this methodological restriction should be abandoned is supported by Bohr's own comment about certain quantum processes "transcending space and time," which, contrary to his other pronouncements, unambiguously testifies to knowledge gained from quantum theory concerning the possible existence of a realm transcending space and time.

Indeed, as Einstein and others have noted, there appears to be a deep and significant connection between certain mathematical objects and physical reality – were that not the case, the whole field of theoretical physics would be without power or purpose in providing an account of the empirical realm. There is ample precedent for entities and procedures that seem purely formal and abstract turning out to have concrete physical relevance. For example, in the words of Freeman Dyson, the mathematicians of the nineteenth century

had discovered that the theory of functions became far deeper and more powerful when it was extended from real to complex numbers. But they always thought of complex numbers as an artificial construction, invented by human mathematicians as a useful and elegant abstraction from real life. It never entered their heads that this artificial number system that they had invented was in fact the ground on which atoms move. They never imagined that nature had got there first. (Dyson, 2009)

2.8 Science as the Endeavor to Understand Reality

As argued in the foregoing, I believe that quantum theory can present us with a new kind of understanding of nature, based on a wholly new kind of model, *if* we listen carefully and open-mindedly to what the formalism is saying. I take such a new understanding of nature afforded by a theory as an "explanation" of the empirical phenomena in the domain of the theory. However, for those who demand that a model be constructed out of actual, "things of the facts" (by this I mean ordinary, causal, "classical" facts as referred to by the oft-used term "local realism"), there can of course be no such "explanation," as nearly a century of determined attempts has revealed. The failure of classical model-making has been well established and has largely been answered by a turn to strict empiricism and even frank instrumentalism by many researchers who assume, with Bohr, that all models must be classical. Empiricist approaches are essentially Bohrian in character, denying that the job of science is to "understand how nature is" and rejecting the whole idea of model construction as a misguided "demand for explanation" that need not be met (cf. Van Fraassen, 1991, p. 372).[29] In this

[29] Moreover, Van Fraassen (1991, p. 24) conflates the possible existence of "randomness" with "no explanation" in passages such as this, addressing specific outcomes or asymmetries with no apparent antecedent cause: "[Pierre] Curie's putative principle [that "an asymmetry can only come from an asymmetry"] betokens only a

perspective, it is seen as virtuous to renounce explanation in science, and a sign of enlightened wisdom to content ourselves with classifying and predicting phenomena. But, as argued above, this position does not follow logically from the failure of inappropriate mechanistic, deterministic, local (classical) models, and it is at odds with arguably the most important and exciting aspect of the scientific mission: the *discovery* of previously unseen and unknown aspects of reality (a case in point being the atom and its constituents). If we reconceptualize the process of modeling in light of quantum theory, perhaps we can find a new and more fruitful means of discovery.

thirst for hidden variables, for hidden structure that will explain, will answer *why?* – and nature may simply reject the question." In this regard, there may not be a causal, determinate, mechanistic account, but that doesn't mean that there can be *no* account of relevant and interesting additional structure, so the pursuing of such an account is not merely evidence of a futile "thirst for hidden variables." For instance, there is no *deterministic* account of how one ground state is selected from among many possible ones in spontaneous symmetry breaking, yet one can certainly give an account of the process of symmetry breaking in terms of an additional structure which sets the stage for the circumstance of symmetry breaking. This point is addressed in Chapter 4.

3

The Original TI

Fundamentals

3.1 Background

In his famous *Lectures*, Richard Feynman said:

Now in the further development of science, we want more than just a formula. First we have an observation, then we have numbers that we measure, then we have a law which summarizes all the numbers. But the real glory of science is that we can find a way of thinking such that the law is evident. (Feynman et al., 1964)

The Transactional Interpretation of quantum mechanics (TIQM) is precisely that "way of thinking such that the law is evident." In this case, the law in question is the Born Rule for the probabilities of outcomes of measurements. In this chapter, we introduce the basics of TIQM (or more concisely, just "TI").

TI was first proposed by John G. Cramer in a series of papers in the 1980s (Cramer, 1980, 1983, 1986, 1988). The 1986 paper presented the key ideas and showed how the interpretation gives rise to a physical basis for the Born Rule which prescribes that the probability of an event is given by the square of the wave function corresponding to that event. TI was originally inspired by the Wheeler–Feynman (WF) time-symmetric theory of classical electrodynamics (Wheeler and Feynman, 1945, 1949). The WF theory is also called the "absorber" theory or the "direct-action" theory, because it abolished the idea of the electromagnetic field as a separate mechanical system and proposed instead that radiation is a direct interaction between emitters and absorbers, without any independently existing field as an intermediary. The interaction is a time-symmetric process, in which a charge emits a field in the form of half-retarded, half-advanced solutions to the wave equation, and the response of absorbers combines with that primary field to create a radiative process that transfers energy from an emitter to an absorber. This process is symbolized in TI by a "handshake." Let's first review the WF proposal, and then we'll see how TI generalizes the idea to the quantum domain.

3.1.1 The Wave Equation

The wave equation for any field relates the spatial variation of the field to its time variation. For a generic massless wave field denoted by Φ, the wave equation in the absence of sources (called the "homogeneous wave equation") has the form

$$\nabla^2 \Phi - \frac{1}{v^2} \frac{\partial^2 \Phi}{\partial t^2} = 0 \tag{3.1}$$

where v is the speed of propagation of the wave.

To take into account a specific source for the field, a "current" J is added to the right-hand side, giving the "inhomogeneous wave equation"

$$\nabla^2 \Phi - \frac{1}{v^2} \frac{\partial^2 \Phi}{\partial t^2} = J. \tag{3.2}$$

A current J can be a point source such as an electron, or a more extended object (charge distribution) capable of coupling to the electromagnetic field. "Coupling" means having the ability or tendency to emit or absorb photons, the quanta of electromagnetic radiation. The ability for a current to couple in this way to the electromagnetic field is indicated by saying that an object has charge.[1]

3.1.2 Coupling and Absorption in TI

I digress briefly here to note that the concept of coupling is important for understanding the process of absorption in TI, which is often misunderstood. Under TI, an "absorber" is an entity that can generate a confirmation wave (CW) in response to an emitted offer wave (OW). (Both these concepts – OW and CW – are defined in Section 3.2.) The generation of a CW needs to be carefully distinguished from "absorption," meaning the absorption of energy, since not all responding absorbers will in fact receive the energy from a given emitter. In general, there will be several or many absorbers sending CW back to an emitter, but only one of them can receive the emitted energy. This is purely a quantum effect, since the original classical WF absorber theory treats energy as a continuous quantity that is distributed to all responding absorbers. It is the quantum level that creates a semantic difficulty in that there are entities (absorbers) that *participate* in the absorption process by generating CW, but do not necessarily end up receiving energy. In everyday terms, these are like sweepstake entrants that are necessary for the game to be played, but who do not win it.

A long-standing objection to the TI picture has been that the circumstances surrounding absorption are not well defined and that "absorber" or "response of the

---

[1] More precisely, "coupling" means that a current has a nonzero amplitude to emit or absorb a photon.

absorber" are primitive concepts. This objection is addressed and resolved in the current approach as follows. TI indeed provides a nonarbitrary (though not deterministic) account for the circumstances surrounding absorption in terms of coupling between fields. In particular, for the electromagnetic field, the basic amplitude for absorber response is simply the elementary charge, e (in natural units). "Response of an absorber" corresponds, in standard quantum field theory, to annihilation of a quantum state – a perfectly well-defined physical process in the relativistic domain. Since this issue pertains to the relativistic elaboration of TI (referred to as "RTI" in the literature), I defer further details to Chapter 5. The fact that objections to TI can be readily resolved at the relativistic level underscores both (1) the ability of TI to accommodate relativity and (2) the necessity to include the relativistic domain to resolve the measurement problem.

3.1.3 Solutions of the Wave Equation

Returning now to consider the wave equation and its possible solutions, we first need to review some basic features of waves. Any generic wave has a wavelength λ and a frequency f; the speed v of the wave is simply their product

$$v = \lambda f. \tag{3.3}$$

It is customary, for notational convenience, to express λ and f in terms of a "wave number" k and an angular frequency ω, respectively, where

$$k = \frac{2\pi}{\lambda} \text{ and} \tag{3.4a}$$

$$\omega = 2\pi f. \tag{3.4b}$$

Thus, the propagation speed of the wave can also be written

$$v = \lambda f = \frac{2\pi}{k} \frac{\omega}{2\pi} = \frac{\omega}{k}. \tag{3.5}$$

The above is termed the "phase velocity"; it specifies the distance traveled by a particular wave crest in unit time (see Figure 3.1).

In the empirical world, we always seem to see waves diverging outward into space from the past to the future (i.e., from earlier times to later times), as shown in Figure 3.2.[2]

---

[2] We should not, however, *equate* the divergence of the wave with the fact that it is a retarded solution. Retarded waves are simply waves that are created at a source and *later* encounter an absorber. Such waves could be in a light pipe or transmission line and do not necessarily show spherical wave divergence.

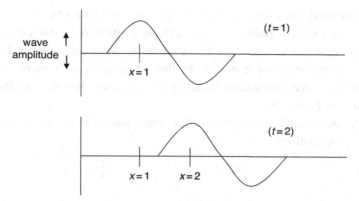

Figure 3.1 The wave crest depicted here travels from $x = 1$ (m) to $x = 2$ (m) in unit time (s), so this wave's phase velocity is 1 m/s. Only one wavelength is shown for simplicity.

Figure 3.2 A falling raindrop creates diverging ripples on the surface of a pond. *Source*: Salvatore Vuono/FreeDigitalPhotos.net.

This type of wave propagation is called "retarded" propagation, and corresponds to a solution to (3.1) of the form[3]

$$\Phi_r(x, t) = e^{\frac{i}{\hbar}\phi} = \exp\left[\frac{i}{\hbar}(kx - \omega t)\right]. \tag{3.6}$$

---

[3] For simplicity, I neglect constant coefficients. In addition, this presentation is a heuristic one in a single spatial dimension x, so it does not reflect the distinction between solutions to the homogeneous and inhomogeneous equations. Strictly speaking, "advanced" and "retarded" solutions apply only to the inhomogeneous wave equation (i.e., the equation with sources) in three spatial dimensions.

We can understand the solution of (3.6) as propagating into the future by seeing that the value of x for any point of constant phase ϕ (such as a wave crest referred to above) increases with increasing time. For example, when $kx = \omega t$, we have $\phi = 0$. In order to keep the phase constant in this expression as t increases, x must increase; so the wave propagates spatially in the same direction as its temporal propagation (see Figure 3.1).

However, an equally (mathematically) valid solution exists in the form of "advanced" propagation:

$$\Phi_a(x, t) = \exp\left[\frac{i}{\hbar}(kx + \omega t)\right]. \tag{3.7}$$

Let's examine the behavior of the phase of this solution as we did for the retarded case. As t increases, the value of the spatial index x must *decrease* to keep the phase constant (i.e., to keep track of the same spot on the wave such as a crest or trough). For the spatial index to increase, the temporal index t must *decrease* (i.e., the wave must propagate "into the past"). If we consider the more realistic three-dimensional situation in which a point source gives rise to the field solutions under consideration, the spatial coordinate x changes to r, which tells us the radial distance from an emitting source.[4]

The retarded solution corresponds to a set of spherical wave fronts (sets of spatial points of constant phase) that diverge with increasing t; that is, r increases with increasing t. In contrast, the advanced solution corresponds to cases in which the spatial and temporal indices increment in opposing directions. This gives us either (1) a set of wave fronts converging onto the source from the past or (2) a set of wave fronts diverging *from* the source *into* the past (depending on which way we choose to orient the "flow" of events with respect to a spacetime diagram). These three-dimensional forms are illustrated in Figure 3.3.

In Figure 3.3, we can think of someone with a stopwatch standing at the origin of a coordinate system and shining a flashlight for a split second when their stopwatch says $t = 0$. As we map the person's experience on a spacetime diagram (consider (a) first), they are at the center of an ever-widening sphere of concentric wave fronts. Because we can't represent three spatial directions plus a time direction on paper, these spherical wave fronts have to be pictured as a series of widening circles (we neglect one spatial dimension so the spheres get flattened to circles). The person's worldline (not pictured) is a straight vertical line through the center of all the circles.

Figure 3.3(a) is the usual "retarded" wave that diverges as the time index increases. In contrast, Figure 3.3(b) shows the "advanced" wave that diverges as

---

[4] In addition, there is a factor of $1/r$, assuming spherical symmetry.

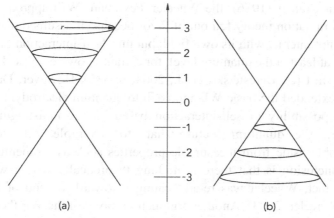

Figure 3.3 Pictured in 2(space)+1(time) dimensions are (a) the retarded wave solution and (b) the advanced wave solution, both with respect to a hypothetical source at the origin (where the light-like diagonal lines cross). The foreshortened circles are actually spherical wave fronts in 3+1 dimensions.

the time index *decreases* – that is, it *propagates into the past.* If we observe the advanced wave from our usual temporal orientation – that is, moving "forward" in time – the advanced wave appears to emerge from all directions and to converge onto the source.

3.1.4 The Wheeler–Feynman Theory

This section gives just a brief conceptual overview of the Wheeler–Feynman theory; further details are provided in Chapter 5. The basic proposal is that all field sources emit half their radiation as retarded and half as advanced; this solution is termed a "time-symmetric" solution. So, in terms of Figure 3.3, the source at the origin (where the light-like diagonals cross) emits equal amounts of both (a) and (b). The retarded component of this field corresponds to the "offer wave" (OW) mentioned in Section 3.1.2. Other charges respond to the emitted time-symmetric field by emitting their own symmetric field, but exactly out of phase with the stimulating field. The advanced component of the response field corresponds to the "confirmation wave" (CW). Wheeler and Feynman showed that the responses of all absorbers, together with the fields from emitters, amounts to the fully retarded field that we seem to see. It also accounts for the loss of energy by a radiating charge, which is problematic in the standard approach.[5]

---

[5] Their result requires a "light-tight box" condition, that is, that any emitted radiation is fully absorbed. However, this condition applies only to the classical version of the theory. We will see in Chapter 5 that the quantum version of the Wheeler–Feynman theory does not really need a "light-tight box."

As noted in Cramer (1986), the Wheeler–Feynman (WF) approach to dealing with classical radiation theory fell out of favor because it assumed that a radiation source could not interact with its own field; but this "self-interaction" was found to be necessary, at least at the quantum level, for certain known empirical effects such as the Lamb shift (cf. Berestetskii et al., 2004, p. 535). However, Davies (1970, 1971, 1972) extended the basic WF approach to quantum electrodynamics, which included the possibility of self-interaction based on the indistinguishability of currents (i.e., the quantum feature that, for example, all electrons are indistinguishable aside from measurable properties such as momentum or spin). The basic conclusion is that there is nothing theoretically wrong with the WF approach. In fact, Wheeler was re-advocating it toward the end of his life (see Wesley and Wheeler, 2003). Another argument in favor of the WF theory is that it resolves long-standing consistency problems of standard quantum field theory. This issue is discussed in Kastner (2015).

3.2 Basic Concepts of TI

Cramer (1986) specified the ways in which TI differed from traditional approaches to interpreting quantum mechanics (in particular, the Copenhagen interpretation) and argued instead for a realist approach in which the theoretical quantum state $|\Psi\rangle$ and its adjoint $\langle\Psi|$ represent real physical entities in a time-symmetric interpretation based on the basic WF formalism. He showed that the Born Rule for calculating the predicted probabilities of observable events arises naturally from the interaction of offer waves (OW, represented mathematically by the usual quantum state or wave function) and confirmation waves (CW, represented by the adjoint quantum state or complex conjugate wave function). In the remainder of this chapter I review the key features of this original proposal.

3.2.1 Emitters and Absorbers

Emitters and absorbers are simply those entities in standard physics that can emit or absorb another quantum. More technically, as addressed briefly in Section 3.1.2, emitters and absorbers are field currents that can couple to other fields, which means that they have an amplitude to emit or absorb quanta of those fields. This is actually the physical meaning of "charge," since (as noted earlier) the basic coupling amplitude for electromagnetism is the elementary charge e. Emission and absorption correspond to creation or annihilation of quantum states, respectively. However, as noted in Section 3.1.2, in the quantum context the term "absorption" gains an ambiguity because it does not necessarily correspond to the reception of a quantum of real energy. Thus, for clarity going forward, we'll restrict the use of "absorption" to the reception of real energy, and use "absorber response" for the generation of a CW that may or may not lead to actual absorption by the entity that generated it.

Examples of emitters or absorbers are electrons, which can serve as emitters or absorbers of photons. However, it's important to note that electrons can only emit or absorb as components of bound states, since a free electron can neither emit nor absorb due to energy and momentum conservation. Thus, in general it's more accurate to refer to atoms and molecules (bound states) as emitters and absorbers. An exception is the case in which a free electron emits and as a result becomes part of a new bound state (as in radiative recombination, the inverse process of the photoelectric effect), which allows it to satisfy the conservation laws.

So, in short, an emitter is a system that can generate an offer wave (OW). An absorber is a system that can respond to an emitted OW with a confirmation wave (CW), whether or not that particular system actually ends up receiving energy. The basic probability of both OW and CW generation is characterized by the square of the charge, or fine structure constant (since these processes must occur together; we'll examine that in further detail in Chapter 5). An additional factor concerns the available energy states, since any such emission or absorption must obey the conservation laws. So, at the micro level, any particular field source can be considered a potential emitter or potential absorber, depending on its available energy states.

3.2.2 Offer Waves and Confirmation Waves

The term "offer wave" (OW) denotes the entity referred to by the usual quantum state $|\Psi\rangle$, which corresponds to the retarded component of the field in the Wheeler–Feynman account. An OW is what is emitted by an emitter (along with the emitter's advanced wave component). A "confirmation wave" (CW) is the advanced component of the response field generated by an absorber and is represented by the dual state or "brac" $\langle\Phi|$ (the state labels are arbitrary here). The CW corresponds to the advanced component of the field in the Wheeler–Feynman account. The process whereby the absorber's advanced field (CW) reinforces the emitter's retarded field, and the remaining advanced component from the emitter and retarded component from the absorber are canceled, is illustrated in Figure 3.4.

Since it is well known that the operation of time reversal takes "kets" into "bracs,"[6] this gives a natural time-symmetric interpretation of the ubiquitous inner product quantities appearing in quantum theory, such as $\langle\Phi|\Psi\rangle$. We come across an inner product form when taking into account the fact that an absorber

---

[6] See, e.g., Sakurai (1984, pp. 273–74).

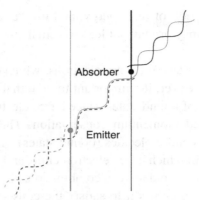

Figure 3.4 The advanced field (dashed line) from the absorber exactly reinforces the retarded field (solid line) between the emitter and absorber and exactly cancels the advanced field from the emitter and the retarded field from the absorber, so all that remains is a fully retarded wave carrying energy from the emitter to the absorber.

corresponding to the property labeled Φ can only absorb (annihilate) that component of any OW encountering it. This can be thought of as "attenuation" of the original OW from the perspective of the absorber corresponding to Φ: the component of the OW in the state labeled Ψ reaching an absorber corresponding to the state labeled Φ will be the projection of $|\Psi\rangle$ onto $|\Phi\rangle$, or $\langle\Phi|\Psi\rangle|\Phi\rangle$.

In the next section I review how the Born Rule, or the probability of an outcome corresponding to the property Φ for a system prepared in state $|\Psi\rangle$: $P(\Phi|\Psi) = |\langle\Phi|\Psi\rangle|^2$, arises naturally in TI.

3.2.3 The Born Rule Is Revealed in TI

Let us consider the general case, in which an emitted OW labeled $|S\rangle$ from a source S of quanta (such as a laser) encounters absorbers labeled by properties A, B, C, D, ... (see Figure 3.5). As described in the previous section, the component absorbed by A is $\langle a|s\rangle|a\rangle$ and the component absorbed by B is $\langle b|s\rangle|b\rangle$, and so on. Each absorber responds with the advanced CW $\langle s|a\rangle\langle a|$, $\langle s|b\rangle\langle b|$, and so on. The product of the OW and CW amplitudes gives the Born Rule for the probability of the outcome, for example, $P(A|S) = \langle s|a\rangle\langle a|s\rangle = |\langle a|s\rangle|^2$. Formally, the interaction of the entire OW ket and CW brac corresponding to each absorber is represented by an outer product, which yields a projection operator multiplied by the Born Rule (Figure 3.5).

The preceding account leads to a weighted set of "competing" possible transactions that we can call "incipient transactions." Note that all possible transactions are associated with projection operators, that is, matching "final" OW

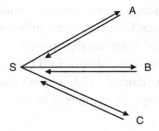

Figure 3.5 An offer wave $|S\rangle$ can be resolved into various components correspond-
ing to the properties of absorbers A, B, C, ... The outer product of a particular OW
component $\langle a|s\rangle|a\rangle$ with its corresponding CW component $\langle s|a\rangle|a\rangle$ yields
a projection operator $|a\rangle\langle a|$ multiplied by the Born Rule, $\langle a|s\rangle\langle s|a\rangle = |\langle a|s\rangle|^2$.

and CW components. Thus, the weighted set of incipient transactions is just von
Neumann's "Process 1," discussed in Chapter 1.[7] To review, von Neumann
proposed that, upon measurement of an observable O with possible values X_i
(these correspond to A, B, C, D, above) on a system prepared in state $|\Psi\rangle$, the
system's state undergoes a change from a "pure state" to a "mixed state," that is,

$$|\Psi\rangle\langle\Psi| \to \sum_i |\langle\Psi|X_i\rangle|^2 |X_i\rangle\langle X_i|. \tag{3.8}$$

The notorious problem with the von Neumann formulation was that there seemed
to be no way to determine when, why, or how the pure state should undergo such a
transformation. If we take into account the physical process of absorption (i.e., state
annihilation), "Process 1" becomes completely non-mysterious. It is just the
process whereby the CW are returned to the emitter from all absorbers capable of
responding, and a set of incipient transactions is established.

The "mixed state" on the right-hand side of (3.8) represents a set of incipient
transactions, of which (in general) only one can be actualized. However, the
presence of absorbers defines unambiguously the basis with respect to which the
offer wave must be decomposed, thus eliminating many of the perplexing
ambiguities often present in discussions of the quantum state (which can
theoretically be expressed in myriad such bases). Here, the "observable" being
measured is the operator defined by the sum of the incipient transactions
represented by $|X_i\rangle\langle X_i|$ in (3.8), where each is multiplied by its associated
eigenvalue (i.e., the "value of the observable" corresponding to that outcome). The
latter is referred to in the literature as the "spectral decomposition" of the
observable.

---

[7] See also Bub (1997, p. 34).

The weighted set of incipient transactions corresponds to a classical probability space in which the weights can be straightforwardly interpreted as the probability that the answer "yes" can be consistently applied to questions such as "Is the system in state X_k?" There is true collapse in TI, in that the property ultimately selected is stochastically actualized with the corresponding probability. This collapse is understood as a type of symmetry breaking; the latter is discussed in detail in Chapter 4.

3.3 "Measurement" Is Well Defined in TI

As is evident in the foregoing, the key advantage of TI over other "collapse"-type interpretations is that the notion of "measurement" is unambiguously defined in physical terms, without appeal to the consciousness of an external observer, and thus without the usual "shifty split" or "Heisenberg cut" which inevitably attends the attempt to specify what counts as "external." In this section, I compare TI's treatment of measurement with competing accounts and see how it provides a solution to the measurement problem, in the sense of making clear at what point a measurement can be said to actually occur.

3.3.1 TI's Advantages over Traditional "Collapse" Interpretations

A system undergoing measurement in TI is actualized in a particular property that can be described in classical terms – that is, it gains a determinate property that could be corroborated through repeated measurements. The account is not relational in that this actualizing event definitely occurs, as an objective matter. It is "contextual" only in the sense that its actualization becomes possible when a particular observable, characterized by the spectral decomposition, is being measured (as opposed to some noncommuting observable).

How is this achieved? Very simply, by taking into account that *absorption is a real physical process*. This is certainly the case in relativistic quantum field theories: one cannot arrive at a correct empirical prediction without taking absorption into account. Indeed, absorption (i.e., annihilation) is a key element of the definition of the field operators used in any calculation of probabilities of empirical events. Such calculations routinely involve taking expectation values in which quanta are created and quanta are destroyed. If such calculations refer to anything physical (the basic realist assumption), both processes are physical processes. However, for the past century or so, interpretations of nonrelativistic quantum theory have completely disregarded the absorption process, granting physicality only to emission processes giving rise to quantum objects that are described by the usual (retarded) quantum states ("kets"). They have thus considered nonrelativistic quantum mechanics – which is just a limiting case of quantum theory – only in a particular form (as applying only to emission) and in isolation from its relativistic application; and this, along with neglect of the absorber theory of fields (i.e., the Wheeler–Feynman direct-action picture), is what has prevented the ability of such interpretations to account for measurement in physical terms.

Specifically, a measurement or determinate event (i.e., it does not have to be a formal "measurement" conducted by an observer) occurs whenever annihilation of one or more real quanta occurs.[8] In terms of relativistic quantum theory, absorption corresponds to the action of annihilation operators on free quanta, just as emission corresponds to the action of creation operators on the vacuum state.[9]

As noted earlier, a common objection to TI is the claim that absorbers are not well defined, but this objection apparently ignores the fact that absorbers *are* well-defined objects throughout physics. If emitters are taken as well defined – that is, if we can assert that it makes ontological sense to say that the entity described by a quantum state is emitted (created) – then one cannot consistently argue that it doesn't make ontological sense to say that the entity described by a quantum state is absorbed (annihilated).

As noted in Section 3.1.2, one source of confusion surrounding TI is that "absorption" is sometimes conflated with "detection" – that is, with empirically detectable transfer of energy from an emitter to an absorber. But (in terms of quantum field theory) an annihilation operator corresponding to property A can act without necessarily resulting in an actualized event A, just as the creation operator corresponding to property B can act without resulting in an actualized event corresponding to property B. For example, the ket $|p\rangle$ can be written in terms of a creation operator as $a^\dagger(p)|0\rangle$, which can be understood as the creation of the *possibility* of property p from the vacuum. Now, recall that a particular momentum state can be written as an infinite sum of all possible position states $|x\rangle$.[10] If a measurement of position is then performed, the quantum will be detected at some position x. What happened to the property p? It was not actualized. Note that the brac $\langle p|$ can be written as $\langle 0|a(p)$. This corresponds to the destruction of *the possibility* of property p, just as $|p\rangle$ corresponds to the creation of the possibility of property p.

---

[8] I add the qualifier "real" because *virtual* (i.e., off-shell) particles do not prompt confirmations. This topic is discussed in Chapter 5.

[9] For example, the action of the creation operator for momentum p on the vacuum state $|0\rangle$ yields the state $|p\rangle$ ($|p\rangle$ is emitted), while the action of the annihilation operator for momentum p on the state $|p\rangle$ yields the vacuum state ($|p\rangle$ is absorbed).

[10] At the relativistic level, one uses instead field operators $\hat{\phi}(x)$; there are no genuine position eigenstates in quantum field theory.

Alternatively, one can create a photon state (offer wave) corresponding to horizontal plane polarization, and then allow it to interact with a polarizer oriented at some angle θ with respect to the horizontal. Absorbers in the polarizer and at a final detector both act on that state to destroy (absorb) the corresponding property – either θ or $\theta + \pi/2$ – by generating corresponding CW, but the photon is only actually detected by one of them (with corresponding Born probabilities). The key point is that *the absorption (annihilation) of the entity described by a quantum state is not the same as empirical detection of an actual quantum.* The identification of quantum states as possibilities is explored further in Chapter 4.

For now, we can also note that the above account provides a nice explanation of "null measurement." Recall that if more than one responds to the emitted OW, we have a competition among the responding absorbers such that only one of them actually receives the quantum. Nevertheless, the non-unitary measurement transition occurs once confirmations are generated, since the transactional process gives rise to a fact of the matter about which of the responding absorbers receives the quantum and which do not. For those that respond but do not receive the quantum, a "null measurement" has occurred. Thus, the transactional picture elegantly accounts for "null measurements" in terms of the non-unitary process involving confirmations, and the fact that the photon can only go to one of the responding micro-absorbers.

In a nutshell, TI treats absorption on the same dynamical footing as emission, providing an unambiguous account of how a "measurement" is finalized, without the infinite regression of apparatus or observers infecting the standard accounts of quantum measurement that neglect absorption. It is also harmonious with relativity (as we will see in Chapter 5) and finds support for its even-handed treatment of emission and absorption in quantum field theory, which treats absorption and emission symmetrically.[11] (Emission can be said to be privileged only insofar as it is the starting point for a transaction; something must be created "before" it is destroyed.) Transactions are irreducibly stochastic processes triggered by absorption events. So in TI, measurements – and any other empirically observable events – are just the results of actualized transactions. There is no need to assign wave functions to macroscopic pointer coordinates, observers, or observer minds; nor, under TI, would this be correct – since an offer wave describes an unactualized possibility while macroscopic objects such as pointers and observers (at least those aspects of them that are accessible empirically) are conglomerates of

---

[11] In this regard, note that the expression for a quantum field operator associated with a particular spacetime point is a sum of creation (emission) and annihilation (absorption) operators (cf. Mandl and Shaw, 1990, p. 44). Absorption is just as important as emission in relativistic theories. It is only in traditional nonrelativistic quantum mechanics interpretations that absorption is ignored; TI remedies that discrepancy. This is not to say that TI finds its best relativistic expression in terms of QFT; the more suitable approach is that of the direct-action theory of Davies (1970, 1971, 1972). This point is discussed in Chapter 5.

actualized events based on completed transactions. We discuss this issue further in Chapter 7.

3.4 TI Sheds Light on Feynman's Account of Quantum Probabilities

I now examine a presentation by Feynman in his famous *Lectures on Physics* (Feynman et al., 1964), vol. 3, in which he explains the rules for calculating probabilities of outcomes by reference to the two-slit experiment (recall Chapter 1). Feynman's presentation, while eminently readable, raises intriguing questions about when or why an experiment is considered "finished," which can be answered in the TI picture. Let's first review his discussion.

3.4.1 Feynman's Discussion of the Two-Slit Experiment

Feynman considers a two-slit experiment with electrons, where there is an option to detect which slit each electron went through by shining a light source on the slits. The basic setup, as presented by Feynman, is reproduced in Figure 3.6.

An electron gun emits electrons that can yield interference patterns at the final screen, detected through varying count rates for each position x on the screen. A light source behind the slitted screen emits photons, which can be scattered by the electron into detectors 1 and 2 corresponding to which slit the electron went through. The higher the photon's frequency, the smaller its wavelength and the more accurate its which-slit detection. More specifically, the sharp measurement consists of aiming the photon precisely at one of the slits so that it has no chance of intercepting an electron going through the other slit. A fuzzier measurement consists of the photon having some uncertainty in its "aim" (momentum direction) so that it has chance of intercepting an electron going through the other slit.

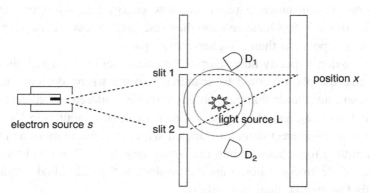

Figure 3.6 The setup for the two-slit experiment with possible "which-slit" detection.

Feynman presents a quantitative analysis of this experiment, which I first review in standard terms and then in terms of TI.

The amplitude for an electron to go from its source s (the electron gun) to slit 1 is $\langle 1|s \rangle$, and similarly the amplitude for an electron to go from s to slit 2 is $\langle 2|s \rangle$. Feynman highlights the "right to left" character of the notation, in which the emitted state is $|s\rangle$ and the projection of $|s\rangle$ onto $|1\rangle$ has the amplitude $\langle 1|s \rangle$ as discussed in Section 3.2.2. Now, in the absence of any detection at the slits, these are just amplitudes, and each is then multiplied by the amplitude to go from either slit to an arbitrary position x on the screen. Thus, the electron amplitudes to go *from* the source, *by way of* slit 1 or 2, *to* position x are:

$$\phi_1 = \langle x|1 \rangle \langle 1|s \rangle \tag{3.9a}$$

$$\phi_2 = \langle x|12 \rangle \langle 2|s \rangle. \tag{3.9b}$$

That is, the amplitude to go from the source to slit 1 (or 2) is multiplied by the amplitude to go from slit 1 to the position x on the screen. Feynman notes here that the rule for calculating the probability of an outcome for intermediate unobserved ("indistinguishable") states is to multiply the amplitudes for each step of the process, and then to add those amplitudes for the overall amplitude of the process. Finally, one squares that amplitude to get the probability that an electron starting out from the source ends up at position x, given that both slits are open.

But we're not done yet. The next step in the analysis is to take into account the emission of a photon from light source L each time an electron goes through the apparatus. The photon has a certain amplitude to be scattered into either detector D_1 or D_2. That amplitude depends on the design of the apparatus and the energy of the photons. For example, in an ideal, sharp measurement, the photon will *only* be scattered into D_1 by an electron state that corresponds to "going through slit 1." But Feynman keeps the analysis general to allow for less precise, or "unsharp," measurements. For instance, a photon of low energy has a longer wavelength, which means that it is less localized and therefore gives a less precise measurement of the electron's position than a higher-energy photon.

Suppose we don't specify how sharp the measurement is, and just allow for the possibility that the photon could be scattered into the wrong detector: that is, we allow a nonzero amplitude that the photon could be scattered by an electron going through slit 1 into D_2, and vice versa. Feynman calls the amplitude for scattering the photon into the correct detector (labeled with the same number as the slit) a, and the amplitude for scattering into the wrong detector b. The amplitudes for the total system of electron + photon are the product of the individual amplitudes, so we have (first in words, then in symbols):

Amplitude for photon to go from L to D_1 and electron to go from s to x by either slit = [(amplitude for electron going from s to slit 1) times (photon "correct" amplitude a) times (amplitude for electron going from slit 1 to x)] plus [(amplitude for electron going from s to slit 2) times (photon "incorrect" amplitude b) times (amplitude for electron going from slit 2 to x)] =

$$\langle x|1\rangle a\langle 1|s\rangle + \langle x|2\rangle b\langle 2|s\rangle = a\phi_1 + b\phi_2 \tag{3.10a}$$

and similarly:

Amplitude for photon to go from L to D_2 and electron to go from s to x by either slit =

$$\langle x|2\rangle a\langle 2|s\rangle + \langle x|1\rangle b\langle 1|s\rangle = a\phi_2 + b\phi_1. \tag{3.10b}$$

Now comes the interesting part. The photon can end up in two "distinguishable" states, either at D_1 or D_2 (where I put scare quotes around "distinguishable" since this is what needs ontological disambiguation). If we want the probability that the electron ends up at x and the photon ends up at either detector, then according to the standard account, because the photon detections are "distinguishable," we square the individual amplitudes applying to each photon detector (Equations (3.10a) and (3.10b)) and then add those squared quantities:

$$P \text{ (electron at } x, \text{photon at } D_1 \text{ or } D_2) = |a\phi_1 + b\phi_2|^2 + |a\phi_2 + b\phi_1|^2. \tag{3.11}$$

Now, let us check that we get an interference pattern if the photon "measurement" is maximally fuzzy; that is, if $a = b$ (in this case it works out that their value must be $1/\sqrt{2}$):

$$P_{\text{fuzzy}}(\text{electron at } x, \text{photon at } D_1 \text{ or } D_2) = \frac{1}{2}\left(|\phi_1 + \phi_2|^2 + |\phi_2 + \phi_1|^2\right)$$

$$= \frac{1}{2}\left(2|\phi_1|^2 + 2|\phi_2|^2 + (\phi_1^*\phi_2 + \phi_2^*\phi_1)\right)$$

$$= |\phi_1|^2 + |\phi_2|^2 + (\phi_1^*\phi_2 + \phi_2^*\phi_1) \tag{3.12}$$

which is the same result as if there were no photon at all, and interference is evident in the cross terms.

On the other hand, for a perfectly sharp measurement, $a = 1$ and $b = 0$, so we get

$$P_{\text{sharp}}(\text{electron at } x, \text{photon at } D_1 \text{ or } D_2) = |\phi_1|^2 + |\phi_2|^2 \tag{3.13}$$

which clearly loses the cross terms and the interference.

I reviewed this discussion by Feynman in the traditional manner because his account of the rules for calculating probabilities raises some interesting questions that are well answered in the TI picture. For example, here is what Feynman says about the difference between the conditions requiring (1) adding individual amplitudes before squaring and (2) squaring individual amplitudes first and then adding them:

Suppose you only want the amplitude that the electron arrives at x, regardless of whether the photon was counted at D_1 or D_2. Should you add the amplitudes [for Equations (3.10)]? No! *You must never add amplitudes for different and distinct final states.* Once the photon is accepted by one of the photon counters, we can always determine which alternative occurred if we want, without any further disturbance to the system.... [D]o not add amplitudes for different final conditions, where by "final" we mean at the moment the probability is desired – that is, when the experiment is "finished." You do add the amplitudes for the different *indistinguishable alternatives* inside the experiment, before the complete process is finished. At the end of the process, you may say that "you don't want to look at the photon." That's your business, but you still do not add the amplitudes. Nature does not know what you are looking at, and she behaves the way she is going to behave whether you bother to take down the data or not. *(Feynman et al., 1964, vol. 3, pp. 3–7; emphases in original)*

But *what* is it that nature is doing that is independent of whether we look or not? What physical circumstance defines when the experiment is "finished"? What makes the two photon states "distinguishable"? What counts as a "disturbance" and what doesn't? These questions are at the very core of the measurement problem and are not answered in the usual pragmatic approaches, which use language like "distinguishable" or "irreversible" without being able to define those conditions in unambiguous physical terms. In particular, according to the usual approach, there is supposedly an ongoing entanglement of the quantum systems with objects in their environment, including measuring apparatus.[12] This is the point of Schrödinger's cat paradox. Despite Feynman's language about nature doing what she does whether or not we are looking, the criteria for when experiments are "finished" inevitably end up referring to the choices of experimenters as to what to measure and/or what can be distinguished *by experimenters*. So, Feynman's obvious (and laudable, in my view) intent to portray the physics as independent of observers and their knowledge sidesteps the fact that the usual account inevitably drags observers back in. (This awkwardness is highlighted by his choice to put "finished" in quotes.) Let's now see now how TI resolves this conundrum.

---

[12] The decoherence program is concerned with showing that this alleged entanglement reduces to an approximately classical world, but as noted in Chapter 1, that program depends on ad hoc assumptions about which part of the universe is the system under study and which is its environment. This makes the account observer-dependent and of course conflicts with Feynman's portrayal of nature as "doing what she does" regardless of our knowledge or choices. In fact, there is no "classical world" in the decoherence approach unless seeds of classicality (localization of systems and/or separability of the Hilbert space) are imposed at the outset, which makes the account circular.

3.4.2 TI as the Ontological Basis for Feynman's Account

First, recall that in TI there is no "measurement" – indeed, no actualized event – *unless confirmation waves (CW) are generated by an absorber.* I illustrate a more precise setup for the experiment in Figure 3.7.

The interaction between the photon and the electron is illustrated schematically in Figure 3.8. The amplitudes *a* and *b* play the part of scattering amplitudes for various incoming and outgoing states of the photon and electron. For *a* and *b* arbitrary, and the light source L aimed at slit 1, we have the possible scattering events shown in Figure 3.8 corresponding to Equation (3.10a).

In calculating the amplitude for a scattering process, one considers all the different ways (up to a given order) that a particular set of events can occur. In this

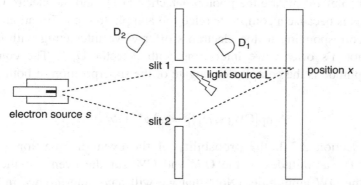

Figure 3.7 The more precise setup showing the light source aimed at slit 1.

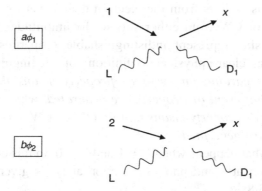

Figure 3.8 The two scattering amplitudes contributing to activation of detector D_1 for the photon for an imperfect measurement. The top diagram corresponds to the correct photon detection and the bottom diagram corresponds to the incorrect photon detection. The electron is represented by the straight line and the photon by the wavy line.

case, we are interested in all the ways that an electron and photon can start out in states $|s\rangle$ and $|L\rangle$, respectively, and end up in states $|x\rangle$ and $|D_1\rangle$, respectively. I suppress the electron's initial state $|s\rangle$ and just indicate whether it has "gone through" slit 1 or 2 by the corresponding number. The electron amplitude for the top diagram is ϕ_1, and for the bottom diagram it is ϕ_2. The photon amplitude for the top diagram is a, and for the bottom diagram is b. The two-particle amplitude for each diagram is the product of the individual particles' amplitudes, and the total amplitude for the scattering process is the sum of the two amplitudes for the diagrams.

Now let's consider the foregoing in the TI picture. The offer wave corresponding to "electron goes through either slit, photon goes to D_1" is a superposition of the two components illustrated in Figure 3.8. The only "distinguishable" events are the photon detection at D_1 and the electron detection at x, and this is because a composite (electron and photon) confirmation wave was generated corresponding to the electron's offer wave interacting with detector x and the photon's offer wave interacting with detector D_1.[13] The confirmation wave's amplitude is the complex conjugate of the superposition of both diagrams, that is,

$$\text{Amp}[\text{CW}(x, D_1)] = a^* \phi_1^* + b^* \phi_2^*.$$

Recalling Section 3.2.3, the probability of the event in question is just the product of the amplitudes of the OW and CW for the event, or the absolute square of the OW amplitude. (Note that we will have interference in this case, due to the fact that b is different from zero.) The reason that we square the sum of the amplitudes for the D_1 photon detection is precisely because *there is a CW from detector D_1* (as well as from the electron detector at x). In contrast, there was never an OW or CW from either slit, so the amplitudes corresponding to "going through a slit" represent indistinguishable properties or occurrences. Thus, we have a clear physical definition of distinguishability in TI: *Distinguishability regarding an event or property means that OW and CW corresponding to that event or property are generated, while indistinguishability regarding event or property means that no OW or CW corresponding to that event or property are generated.*

Now, let us see what happens when $a = 1$ and $b = 0$. In this case, the amplitude for electron detection at x and photon detection at D_1 is given only by the top diagram in Figure 3.8.

---

[13] This treatment of the electron is schematic and approximate, since at the relativistic level, fermionic fields such as electrons participate in transactions indirectly, by way of bosonic offers and confirmations. This issue is discussed in Chapter 6. However, these details do not affect the analysis here.

There are CW generated at x and at D_1, so we square the amplitude for the diagram:

$$P(a = 1, \text{electron at } x, \text{photon at } D_1) = |\phi_1|^2 = \langle x|1\rangle\langle 1|s\rangle\langle s|1\rangle\langle 1|x\rangle$$

which is the probability distribution for an electron going through slit 1 (as if slit 2 were closed). Interference is lost in this case, because the only available transactions (i.e., OW with CW responses) are those projecting the total system onto *either* D_1 and x by way of slit 1 *or* D_2 and x by way of slit 2. Even though there is still no CW from either slit, the sharp correlation between the photon and the electron, disallowing any transaction for the photon ending up at the "wrong" detector, makes the slits distinguishable. In effect, due to the sharp correlation, the photon detectors act as proxies for the electron slits. All CWs for the total system are "which-slit" CWs.

The physical content behind the above probabilities is that the squared quantities represent confirmations of the corresponding offer wave components (recall Section 3.2.3). But it is important to keep in mind that the confirmation of an OW does not necessarily mean that the event or property corresponding to that particular OW is actualized; rather, it means that *there is a determinate fact of the matter* as to whether that property or another property corresponding to the same observable is actualized. In other terms, it means that the usual classical rules of probability apply to the situation – that is, the "cat is either alive or dead," in contrast to the fuzzy quantum logic of saying "the cat is in a superposition of alive and dead."

Note that this account obviates the need to refer to observer-dependent criteria such as (in Feynman's phrasing) "the moment the probability is desired." It allows us to take away the scare quotes from Feynman's reference to the experiment being "finished"; the experiment is unambiguously finished when a confirmation is generated that allows us to apply the rules of classical probability, that is, to say that there is a definite fact of the matter about the whereabouts of both particles in the experiment. Those rules apply not because someone desired a probability, but because there was a *physical process* in play: a confirmation, which brought about a determinate event. The confirmation is what creates the "disturbance" that disrupts the quantum superposition. TI allows us to define what "disturbance" really means, in concrete physical terms.

One last detail needs to be addressed: the reader may worry about the apparent asymmetry of having to choose a slit at which to aim the photon, which in this discussion was arbitrarily chosen as slit 1. To see that this is not a problem (i.e., no generality is lost), let's return to Equation (3.10b) for the amplitude of an electron at x and a photon at D_2:

$$\langle x|2\rangle a\langle 2|s\rangle + \langle x|1\rangle b\langle 1|s\rangle = a\phi_2 + b\phi_1. \tag{3.10b}$$

Note that this holds regardless of which slit is targeted by the photon source. If D_2 is the "null" detector – that is, a photon is detected there only if it failed to interact with an election – a is still the amplitude that an electron going through slit 2 will correctly result in the photon being detected at D_2. This is because an electron going through slit 2 has very little chance of scattering the photon aimed at slit 1 (where that chance depends on the magnitude of the "error" amplitude b) and so it will just continue on to D_2 "by default." The photon won't be scattered into D_2 by the electron, but what matters for the calculation is not *how* the photon gets to its destination but rather the amplitude that it will do so, and the correspondence of that amplitude with the state of the electron. Even if the photon is aimed at slit 1, it correctly gets to D_2 for an electron state corresponding to slit 2. On the other hand, if the photon ends up at D_1 for an electron state corresponding to slit 2, it's only because the measurement is a "fuzzy" one in which the photon is poorly localized and therefore has a chance of being scattered into D_1 even by an electron going through the "wrong" slit.

In the next chapter, we consider a new ontological paradigm that seems inevitably implied by a fully realist interpretation not only of TI itself but of quantum theory in general. This ontological aspect is described by the term "Possibilist Transactional Interpretation" (PTI).

4

The New TI

Possibilist Transactional Interpretation

4.1 Why PTI?

The 1986 version of TI faced some difficulties: (1) the interpretation of multiparticle offer and confirmation waves, (2) the nature of the process leading to an actualized transaction, and (3) the apparent possibility of causal loops leading to either inconsistency or inconclusive quantitative predictions.[1] The issue to which (1) refers is the following: in mathematical terms, multiparticle states for N particles are actually elements of a $3N$-dimensional space.[2] For example, a general quantum state for two particles A and B contains components A_x, A_y, A_z, B_x, B_y, B_z. So a realist interpretation of such states cannot claim that their offer and confirmation waves exist entirely in ordinary space, but instead must allow for their existence in a quantum-level, extra-empirical space.[3] The updated version of TI presented herein, possibilist TI or PTI, addresses this by proposing that offer and confirmation waves represent dynamically efficacious *possibilities* whose collective structure constitutes just such a quantum-level space, which (anticipating the relativistic domain) I will call the *quantum substratum*.[4] This is a form of structural realism, since I do not claim to know the material nature (if any) of these possibilities, but rather claim that the formal structure of the theory reflects an existing structure in the real world, albeit an extra-empirical one.[5]

Concerning (2), the earlier version of TI referred to an "echoing" process which, given the higher-dimensional entities involved, cannot be thought of as a process

---

[1] I refer here to Maudlin (2002, pp. 199–200) and Berkovitz (2002, 2008).

[2] I use the term "particle" here as a convenience because that is the usual language used in this context. However, the "particles" involved should not be thought of as localized, classical particles. They are quanta or bound states of quanta of one or more fields.

[3] In referring to certain entities as "living" in a space, I do not mean to imply that such a space necessarily exists independently as a substance. That is, this locution is not meant to endorse substantivalism about such a space. It just refers to the mathematical characteristics of the manifold of the entities in question.

[4] At the relativistic level, where particles are being created and destroyed and therefore particle numbers are changing, the relevant "higher space" is described by Fock space, the relativistic extension of Hilbert space.

[5] This ontology is discussed in a collaborative work: Kastner et al. (2018).

taking place within the spacetime manifold. Since there is no causal (in the sense of deterministic) way to account for the actualization of one transaction out of several incipient ones, such an actualization is irreducibly stochastic in a way that is not compatible with any causal process within the confines of ordinary spacetime. PTI takes the actualization of a particular transaction (from among a collection of N incipient ones) as an extra-spatiotemporal process, more akin to spontaneous symmetry breaking (SSB) than to a back-and-forth dynamical process within spacetime such as the "echoing" of Cramer's original TI. This suggestion will be discussed further below. With regard to (3), issues involving causal loops or possible deviations from the predictions of quantum mechanics are fully remedied at the relativistic level, as we will see in later chapters.

4.2 Basic Concepts of PTI

In this section, we consider the defining characteristics of PTI.

4.2.1 *Offer and Confirmation Waves Are Physically Real,*
but Subempirical, Possibilities

OW and CW (see Chapter 3) are interpreted ontologically in PTI as *physically real possibilities*. In this context, "real" means physically efficacious but not necessarily *actualized*. (This distinction is elaborated in detail in Chapter 7.) Again, think of the submerged portion of the iceberg in Figure 1.1: from the vantage point of the deck of a ship (representing the empirical realm), we cannot see the submerged portion, but it certainly supports the visible portion and therefore cannot be dismissed as "abstract" or "unreal." In Bohr's words (recall Chapter 2), these entities "transcend the spacetime construct"; however, rather than dismiss them as Bohr did, I allow that they are physically real, even if subempirical.

OW and CW are necessary but not sufficient conditions for an actualized event. The remaining necessary condition for an actualized event is that one particular transaction be actualized from a set of N incipient ones, where N labels the number of absorbers returning CW to the emitter. The adjective "real" thus designates a broader ontological status than "actual"; that is, the set of actual entities is a subset of all real entities.[6]

---

[6] This is roughly analogous to Lewis' treatment, except for the fact that he considers the "actual" designation as merely indexical. In PTI, an actual event has a different ontological status from a possible (real) event in that the former exists in spacetime (the "tip of the iceberg"), whereas the latter exists in the quantum substratum (the submerged portion of the iceberg).

4.2.2 *Emission and Absorption of Quanta Occur in the Quantum Substratum*

Emissions of offer waves and responding confirmation waves are primary dynamical processes that take place in the quantum substratum. Metaphorically speaking, the quantum substratum corresponds to the submerged portion of the iceberg and the ocean in Figure 1.1 (the ocean can be thought of as the vacuum state). More precisely, the quantum substratum is the manifold mathematically described by Hilbert space,[7] the domain of subempirical (meaning not directly observable) quantum theoretical objects such as the entities described by quantum states.

If the idea of viewing the realm described by Hilbert space as real (what philosophers term "reifying" Hilbert space) seems strange, it needs to be kept in mind that many of the quantum objects described mathematically by Hilbert space quantities and routinely assumed as existing *somewhere,* cannot be thought of as existing within a spacetime manifold – in the sense of being localized at a point or within any spacetime region. For example, the ubiquitous "vacuum state" or ground state in the energy representation $|0\rangle$ has no spacetime arguments (i.e., it is not a function of x, y, z, t) and cannot be considered to exist in any well-defined region within spacetime. (For a technical account of why this is so, a good place to start is Redhead (1995).) So, in a pragmatic sense, physicists already take objects described by Hilbert space quantities as real. When pressed, they might respond that "well, of course it's not *real*, because it does not exist in spacetime," but this merely expresses the conventional, often unexamined definition of "real" that insists: "real = existing in spacetime," which is precisely what I claim needs to be questioned. The point is that these objects are assumed to be physically efficacious: they are acted upon by other physically efficacious entities, and they can give rise to concrete observations (e.g., measurement outcomes). Thus, physicists already view them as essentially real, even though they do not exist in spacetime.[8]

It is important to note that emitters and absorbers are "only" quantum possibilities themselves – that is, field excitations that couple to other fields or sets of bound field excitations such as atoms and molecules. A macroscopic emitter or absorber (e.g., a detector) comprises a large number of such micro-emitters and micro-absorbers. Its observable aspect is a network of linked actualized transactions (this notion of linked transactions will be further clarified in Chapters 7 and 8). A macroscopic object termed an "emitter" is just a type of object[9] with a high probability of generating fields that can couple to other fields

---

[7] Technically, this is really Fock space, the relativistic extension of Hilbert space (see note 4).

[8] Another approach consists in taking quantum objects as epistemic (i.e., referring primarily to the knowledge and/or intentions – for example, of what physical quantity to measure – of an observer). This approach has been critiqued earlier in Chapters 1 and 2.

[9] Fields (e.g., 2011) is skeptical of the idea that such objects can be considered "emergent" or well-defined absent a classical distinction between objects, a distinction which requires an observer. However, for present purposes,

(i.e., microscopic emitters and absorbers). An example is a laser. The atoms that emit the photons are "just" conglomerations of field excitations and exist in the quantum substratum, not in spacetime.[10]

If the previous sentence seems surprising, remember that Ernst Mach railed against the idea that atoms were real because they were not directly observable. We have become so accustomed to the concept of atoms that we have forgotten that they are not directly observable: we never really "see" an atom. We can image small numbers of atoms through interactions with devices such as a scanning tunneling electron microscope, but those images are created from transactions between the apparatus and the imaged sample. The transactions result in a measurable current, and there are variations in this current (fewer or more transactions) depending on how many or few atoms comprise a given portion of the sample. The changes in current are rastered onto a two-dimensional surface to yield a kind of "image" of the scanned surface, with regions corresponding to larger currents being identified with atoms. Thus we are not really "seeing atoms"; we are seeing a representation of changes in the transacted current due to interactions between an electron offer wave current and atomic absorbers.

Returning to our basic macroscopic emitter, the apparent solidity of the electrode composed of the atoms is based on the mutual binding among the atoms (through interatomic forces) and on transactions between the atoms and other absorbers (e.g., those in our hands or eyes). This transactional basis of sensory perception is illustrated in Figure 4.1, which shows a man viewing and touching a table (offer waves indicated in black and confirmation waves in gray; remember, though, that these are not entities propagating in spacetime; they are extra-empirical). I will return in Chapter 7 to the epistemic (knowledge-based) implications of this account with reference to the enigmatic and elusive "table-in-itself" discussed at length in Bertrand Russell's *The Problems of Philosophy* (1959, chapter 1).

4.2.3 Incipient Transactions Are Established through OW–CW Encounters in the Quantum Substratum

In a generalization of the Wheeler–Feynman approach discussed in Chapter 3 (on which the original TI was based), PTI (as well as TI) proposes that an absorber

---

I don't need to specify where the "macroscopic emitter" ends and (say) the air surrounding it begins. When humans manipulate macroscopic emitters, they are interacting with a physical entity with a high probability of emitting offer waves, and that is all that is required for this account. That is, the physical entities comprising the "air" portion have a drastically lower probability of emitting offer waves than the physical entities comprising the "laser" portion.

[10] Technically, the fermionic source of a bosonic field (like the electromagnetic field) is a current, which is proportional to the square of the usual quantum state. This is a feature of relativistic quantum mechanics and takes into account that when a photon state is emitted, the original incoming emitter state is modified and becomes an outgoing state.

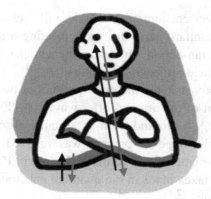

Figure 4.1 Macroscopic objects are perceived via transactions between offer waves emitted by components of the object and confirmations generated by absorbers in our sense organs.

interacts with an offer wave, by generating a confirmation wave.[11] This process can be viewed as a generalization of Newton's third law of motion, which observes that a mass acted upon by a force F exerts an equal and opposite force −**F**. In general, there will be more than one absorber A_i ($i = \{1, N\}$) for an emitted offer wave $|\Psi\rangle$, and in such cases the latter is then projected into components corresponding to the capabilities of each absorber. Formally,

$$|\Psi\rangle = \sum_{i=1}^{N} |A_i\rangle\langle A_i|\Psi\rangle = \sum_{i=1}^{N} \langle A_i|\Psi\rangle|A_i\rangle \qquad (4.1)$$

which reflects the usual projection of a given state $|\Psi\rangle$ onto a particular basis. Thus, PTI provides a physical referent for a common mathematical expression (4.1) in the theory. This interpretation removes the arbitrariness of basis often associated with quantum states. That apparent arbitrariness arises because a crucial aspect of the mathematical formalism has remained physically unapplied in standard accounts of quantum mechanics. Here, we see that the appropriate basis is *physically* determined by the availability of a set of absorbers.

The arbitrariness of basis due to neglecting absorption is analogous to the underdetermination of the force that will be experienced by an object O moving at speed s toward another object O′ when the speed of O′ has not been specified (thus leaving their relative speed unspecified). Just as in classical situations involving Newton's third law, dynamical interactions take place in encounters between OW

---

[11] Taking into account the relativistic domain, one should say that an absorber has an *amplitude* to emit a confirmation wave (just as an emitter has an amplitude to emit an offer wave). This is explored further in Chapter 5. In fact, at the relativistic level, we can see in the basic connection – the time-symmetric propagator – the fundamental origin of Newton's third law.

and CW. A falling object encountering a table will feel a responding force and undergo compression; similarly, an offer wave meeting a confirmation wave will precipitate an incipient transaction which may be actualized. Such encounters are represented by the weighted projection operators $|\langle A_i|\Psi\rangle|^2 |A_i\rangle\langle A_i|$. As noted in the previous chapter, this expression is the outer product of two factors: the projected or attenuated component of the original offer wave, $\langle A_i|\Psi\rangle|A_i\rangle$, and the resulting confirmation wave $\langle \Psi|A_i\rangle\langle A_i|$ to and from a particular absorber.

In the original TI, this type of process was referred to as taking place in "pseudotime," where the latter was a heuristic device. In PTI, this process is fully extra-spatiotemporal; it takes place in the quantum substratum, whose structure is described by Hilbert space.[12]

4.2.4 Spacetime Is the Set of Actualized Transactions

In this interpretation, spacetime is no more – and no less – than the set of actualized transactions. Thus, actualizations of transactions based on OW and CW interactions give rise to the set of related events comprising the spacetime theater. In an actualized transaction, the emission event defines the past and the absorption event defines the future (or, at least, an event at a later time). That is, *the spacetime structure itself supervenes on actualized transactions; there is no "spacetime" without actualized transactions.* The apparent four-dimensional spacetime universe is not something "already there"; rather, it crystallizes from an indeterminate (but real) substratum of dynamical possibility. Thus, spacetime "grows," but not in the usual "growing universe" sense wherein an advancing "now" proceeds from present to future; rather, events arise from a set of dimensions (the Hilbert or Fock space manifold) outside spacetime. In fact, it is the past that "grows" and is extruded from the present; in PTI there is no actualized future. This picture is explored further in Chapter 8.

The apparent temporal asymmetry we observe in the universe is, in part, attributable to the inevitable fact that creation (i.e., emission of a quantum state or OW) necessarily precedes annihilation (generation of a CW); one cannot annihilate unless there is *something already there* to annihilate. This fact is reflected in quantum field theory in the asymmetry in the action of creation and annihilation operators on the vacuum state $|0\rangle$ (which designates that no quanta are present). If you try to operate on the vacuum state with the annihilation operator $\mathbf{a}$, you end up with literally nothing; not even a vacuum state! That is, $\mathbf{a}|0\rangle = 0$. In contrast, you

---

[12] When an absorber generates a CW, it also emits a matching OW (proportional to the ket $|A_i\rangle$), just as in the original TI. Likewise, the emitter also emits an advanced component $\langle \Psi_J|$. However, these residual components mutually cancel. Details are given in the next chapter.

Figure 4.2 A geode is a roughly spherical pocket of crystals growing in a shell of amorphous material. It is built up through mineral deposits in water flowing through lava bubbles and other hollow structures.
Source: https://commons.wikimedia.org/wiki/File:Geode_angle_300x267.jpg.

Figure 4.3 The set of events in spacetime emerges from a pre-spacetime realm of indeterminate possibility, as the inner-ordered, crystalline structure of a geode arises within an outer shell of amorphous mineral. Pictured is Javier Garcia Guinea inside the huge geode he discovered in Almeria, Spain.
Source: Private collection of J. Garcia Guinea, 2002; used with permission.

can act on any state, including the vacuum state, with the creation operator $\mathbf{a}^\dagger$ and still have a well-defined state.

Thus, spacetime arises from beyond itself, from roots of possibility in the quantum substratum. If this picture seems strange or hard to visualize, it can be considered roughly analogous to the formation of a geode (see Figures 4.2 and 4.3). Strictly speaking, a geode forms through the depositing of minerals from surrounding water

Figure 4.4 The circular structure does not exist without the people comprising it. Hans Thoma, *Der Kinderreigen*, 1884.
Source: From www.kunsthalle-karlsruhe.de/kunstwerke/Hans-Thoma/Kinderreigen/ C014 CFF04822A1BCC98882A4CED36A31/.

into a hollow bubble of lava. But if the source of the mineral deposits were in the lava "shell," then the analogy would be closer.

If we think of the geode formation in this latter way, it is an outer "shell" of possibilities which surrounds and gives rise to the crystallized events of the spacetime theater.

More precisely, if the crystals are gradually built up from just inside the shell, that inner layer of shell represents the present, or "now," as experienced by an observer whose sense organs are absorbers on the "receiving" end of a transaction (as in Figure 4.1). The crystalline structure growing toward the center of the geode interior represents the actualized past that continually grows from the now. The outer amorphous shell represents physically real but subempirical content outside this spatial realm in a "higher space" of possibilities.[13] I discuss this metaphysical picture in more detail in Chapters 7 and 8.

At this point, I should touch base with the philosophical terminology for the view of spacetime presented in the preceding section: it is known as *relationalism* or *antisubstantivalism*. This is the view that spacetime does not exist as a substance or as a background "container" for events. Instead, the term "spacetime" describes the structured set of events themselves. This view can be illustrated by reference to Figure 4.4, which shows a group of people forming a circle. We return to this topic in Chapter 8.

---

[13] Technically, the shell represents entities in Hilbert space (or Fock space in the relativistic domain, for noninteracting fields).

A circular structure exists, but it exists only by virtue of the people comprising it. In the same way, according to relationalism, spacetime exists only by virtue of the events comprising it.

4.3 Addressing Some Concerns

Let us now return to concerns (2) and (3) in a little more detail. (Concern (1) is immediately resolved in PTI by positing that quantum state vectors or wave functions represent multidimensional possibilities whose realm is the quantum substratum, not ordinary spacetime.)

4.3.1 How a Transaction Forms

Recall that the subject of concern (2) is that the "pseudotime" process of the original TI does not seem to fully account for why or how a particular transaction is actualized while others are not. If we take the domain of transaction formation as the quantum substratum rather than spacetime, then an account cannot be given in terms of any causal process *within* spacetime in the usual sense – that is, along or within light cones, since the latter are confined to spacetime. Instead, we need to turn to a similar situation in physics in which there are apparently many possibilities but only one is realized: spontaneous symmetry breaking (SSB).

In SSB, the governing theory for the phenomena under study specifies a symmetric situation, illustrated schematically in Figure 4.5. A component of the theory (e.g., a field) undergoes a transformation in which a multiplicity of states or outcomes is possible, none of which can be "picked out" by anything in the theory as the realized state or outcome.

A specific example of this phenomenon occurs in the "Higgs mechanism,"[14] in what is termed the "Standard Model" of elementary particle theory. According to this widely accepted model pioneered by Steven Weinberg and Abdus Salam, the quanta of some force-carrying fields acquire a mass by way of a process in which the ground (vacuum) state of the field undergoes the kind of transformation conceptually depicted above. What was a single vacuum state of the field acquires what is termed a "degeneracy" – that is, many possible ground states (in fact, an infinite number of them). This situation is illustrated in Figure 4.6; the symmetry breaking occurs through what is called a "Mexican hat" potential due to its shape. The original ground state becomes unstable and corresponds to the crown of the "hat"; the infinite set of ground states is found all around the ring at the lowest point. The theory does not provide any way of deciding which of these many

---

[14] The idea was actually arrived at independently in 1964 by Peter Higgs; Robert Brout and Francois Englert; and Gerald Guralnik, C. R. Hagen, and Tom Kibble.

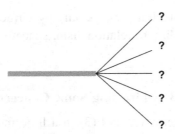

Figure 4.5 Spontaneous symmetry breaking: a transformation of a theory component in which a multiplicity of states or outcomes is possible, none of which can be "picked out" by anything in the theory as the realized state or outcome.

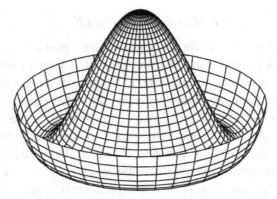

Figure 4.6 The "Mexican hat" potential which creates an infinite number of possible ground states in the Higgs et al. mechanism.
Source: https://commons.wikimedia.org/wiki/File:Mexican_hat_potential_polar.svg.

ground states is realized. But, according to the theory, the fact that the quanta in question have a nonzero mass indicates that one has been realized.

4.3.2 Curie's Principle and Curie's Extended Principle

The situation just described seems to run afoul of a philosophical doctrine termed "Curie's principle" in honor of Pierre Curie, who championed it. (The principle is actually a version of Leibniz' "principle of sufficient reason" (PSR), which states that any event occurs for a reason or cause which specifies or determines that event, as opposed to some other event. The PSR implies that, absent such a reason or cause, the event in question will not occur.)[15]

---

[15] Referring to something as a "philosophical doctrine" simply means that it is presumed to be true on the basis of certain metaphysical or epistemological beliefs or principles. Modern physical theory could be taken as indicating that the principle of sufficient reason may not be applicable to the physical world, however compelling it may seem to those who have championed it.

Figure 4.7 A political cartoon (ca. 1900) satirizing US Congress's inability to choose between a canal through Panama or Nicaragua, by reference to Buridan's ass.
Source: William Allen Rogers, *New York Herald* (Credit: The Granger Collection, NY).

Curie's principle states that an asymmetric result (i.e., the choice of one outcome among many equally possible ones) requires an asymmetric cause. That is, it holds that there can be no sound basis for saying that one of the outcomes "just happens"; one must be able to point to a definite reason for that outcome (the reason being the asymmetric cause). This principle is illustrated by a humorous paradox, "Buridan's ass," named after the French philosopher and determinist Jean Buridan (whom the paradox satirizes), in which a hungry donkey is placed between two equally distant, identical bundles of hay (see Figure 4.7). According to an implicit version of Curie's principle,[16] the donkey will starve to death because it has *no reason* to choose one pile of hay over the other. Of course, our common sense tells us that the donkey will find a way to begin eating hay, even though one can provide no reason for it (hence the paradox). Similarly, in SSB, the field in question arrives in a particular ground state though no specific cause for that choice can be identified. If we take Curie's principle to be applicable to the above, then it appears that nature simply violates the principle (as does a hungry donkey).[17]

[16] Buridan was satirizing the doctrine of moral determinism, which views a person's moral actions and choices as fully determined by past events.

[17] Is there a volitional basis for actualization? Buridan's ass is hungry, so he chooses to eat one of the piles of hay, even if there is no "reason" for it. Does nature then express a certain volitional capacity? Or, put another way, could such an uncaused "choice" be seen as evidence of the creativity of nature?

There is another way of looking at this situation, described by Stewart and Golubitsky (1992). These authors point out that nature seems to be replete with symmetries that are spontaneously "broken," similar to the way in which the symmetry of the vacuum state is broken by the Higgs et al. mechanism. In general, a symmetrical system may, under certain circumstances, be capable of occupying any one of a set of symmetrically related states, with no particular state being privileged; thus, the particular state in which it happens to be found is arbitrary. Stewart and Golubitsky therefore suggest that nature conforms to a weakened version of Curie's principle, which they call the "extended Curie's principle": "physically realizable states of a symmetric system come in bunches, related to each other by symmetry" or, alternatively, "a symmetric cause produces one from a symmetrically related *set* of effects" (Stewart and Golubitsky, 1992, p. 60, emphasis in original). Technically, the "bunches" are subgroups of the original symmetry group which has been "broken" by the dynamical situation under consideration.

As noted by Stewart and Golubitsky, a famous illustration of symmetry breaking appears in the iconic 1957 photograph of the splash of a milk droplet by high-speed photography pioneer Harold Edgerton. I reproduce it here in Figure 4.8. The authors point out that the pool of milk and the droplet both have circular symmetry, but the "crown" shape of the splash does not – it has the lesser symmetry of a 24-sided polygon. This happens because the ring of milk that rises in the splash reaches an unstable point – a point where the sheet of liquid cannot become any thinner – and "buckles" into discrete clumps (the laws of fluid dynamics must be used to predict that there are 24 clumps). But the locations of the clumps are arbitrary; that is, the clump appearing just beneath the white droplet above the crown could just as well have been a few degrees to the left (with all the other clumps being shifted by the same amount). There is thus an infinite number of such crowns possible, but only one of them is realized in any particular splash.

Thus the authors point out that, while the mathematics describing a particular situation may provide for a large, even infinite, number of possible states for a system to occupy, in the actual world only one of these states can be realized. They put it this way:

A buckling sphere can't buckle into two shapes at the same time. So, while the full potentiality of possible states retains complete symmetry, what we observe seems to break it. A coin has two symmetrically related sides, but when you toss it it has to end up either heads or tails: not both. Flipping the coin breaks its flip symmetry: *the actual breaks the symmetry of the potential.* (Stewart and Golubitsky, 1992, p. 60)

I have emphasized the last sentence because it expresses the same deep principle underlying the PTI picture: mathematical descriptions of nature, with their high

Figure 4.8 Milk drop coronet.
Source: Harold E. Edgerton, "Milk-Drop Coronet," 1957. © 2010 Massachusetts Institute of Technology. Courtesy of MIT Museum.

degree of symmetry, in general describe a *set of possibilities* rather than a specific state of affairs. Nevertheless, the astute reader may well raise the following question: But isn't it the case that, in the classical domain, we can always find some external influence, however small, that *caused* the system to end up in one particular state as opposed to some other possible state? This would seem to apply, for example, in classical chaotic systems such as the double pendulum (see Figure 4.9). For large initial momentum, such a system's set of possible trajectories encounters "bifurcation points" (essentially, "forks in the road") in which a specific choice of trajectory is sensitive to perturbations down to the Planck scale (i.e., random quantum fluctuations).

The authors address this, at least in part, as follows:

[W]e said that *mathematically* the laws that apply to symmetric systems can sometimes predict not just a single effect, but a whole set of symmetrically related effects. However, Mother Nature has to *choose* which of those effects she wants to implement.

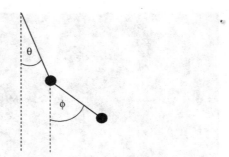

Figure 4.9 A double pendulum, whose classically described motion encounters bifurcation points.

How does she choose?

The answer seems to be: imperfections. Nature is never perfectly symmetric. Nature's circles always have tiny dents and bumps. There are always tiny fluctuations, such as the thermal vibration of molecules. These tiny imperfections load nature's dice in favor of one or the other of the set of possible effects that the mathematics of perfect symmetry considers to be equally possible. (Stewart and Golubitsky, 1992, p. 15)

Thus, the apparent answer of the authors to the question of what causes the system to end up in a particular state is that the cause is found *outside* the mathematical formulation of the set of possible solutions for the system. If we return to the case of spontaneous symmetry breaking in the Standard Model, clearly we are dealing with symmetry breaking in a purely quantum context: the system comprises the vacuum and the Higgs field, purely quantum entities. If we try to find a specific cause – however fleeting and random – for the choice of one of the infinite set of possible vacuum states, we have to either suppose that it stems from the fundamentally indeterministic quantum character of the vacuum or postulate fluctuations in some deeper realm that lies outside any current theory. Alternatively, we might suppose that Nature simply does not respect PRS and that it is possible for Nature itself to "choose" one outcome. The point is still that "the actual breaks the symmetry of the potential," however this is accomplished. The only alternative is to postulate that SSB in the Standard Model requires a "many worlds interpretation," in which SSB gives rise to many possible worlds, each with a different vacuum state. But this is certainly not the usual approach, which simply assumes that the actual universe corresponds to one particular vacuum state.

The foregoing account of spontaneous symmetry breaking can be consistently extended to PTI's account of the realized notion of one particular transaction out of several, or even many, incipient ones. The mathematics describing the situation provides us with a set of possible states of the system, but only one of those can be realized. The new feature appearing in the PTI account is that this set of possible outcomes is weighted by the square of the probability amplitude for that outcome. So the proposed interpretation extends the basic principle of spontaneous symmetry breaking into a weighted type of symmetry breaking over a set of possible states: they certainly cannot all be realized (just as in the case of classical symmetry breaking), so the natural interpretation of the weight of a possible state is as a physical propensity, corresponding to an objective probability of the actualization of the state in question. If we like, we can call this a "weighted symmetry breaking."

Again, note that the establishment of a set of incipient transactions does not require us to adopt a "many worlds interpretation," any more than the "Mexican hat" potential establishing an infinite set of possible ground states requires an "infinitely many worlds interpretation" in the Standard Model. We simply infer that one of the set of possibilities is actualized; Hilbert space describes possibilities and their interactions, while spacetime is the arena of actualized transactions. In the words of Stewart and Golubitsky, "the actual breaks the symmetry of the potential."[18]

4.4 "Transaction" Is Not Equivalent to "Trajectory"

In PTI, a transaction is not equivalent to the establishment of a classical spacetime trajectory, that is, a determinate path from one spacetime point to another. For example, in the two-slit experiment discussed in Chapter 1, with both slits open, a transaction transferring momentum from the emitter to a particular absorber X on the final screen does not establish a particular spacetime path. It retains the wavelike characteristic of a nonlocalized phenomenon in that the quantum (say, a photon) went through both slits.[19] This feature is best understood in the Feynman "sum-over-paths" approach to propagation in quantum mechanics, which I will now briefly review.

4.4.1 Review: Feynman "Sum over Paths"

Nonrelativistic quantum theory is usually formulated in terms of Schrödinger's equation for the propagation of a "wave function," $\Psi(x)$, which is a particular solution to this equation. The wave function is a type of probability amplitude as

---

[18] The issue of symmetry breaking is relevant even in the context of putative deterministic laws, as pointed out by T. M. Mueller (2015). Mueller provides a study of nineteenth-century French philosopher/mathematician Joseph Boussinesq, who noted that some differential equations have more than one solution, and argued that this situation was relevant to the question of free will.
[19] More precisely, the photon interacted with the quanta comprising the boundaries of both slits.

discussed in Chapter 1 (specifically, an amplitude for a quantum to be found at position x if it has been prepared in the state $|\Psi\rangle$; technically, $\Psi(x) = \langle x|\Psi\rangle$). Richard Feynman formulated another approach to this probability amplitude (applicable also in the relativistic regime) by imagining a quantum as an entity that gets from one point to another by taking all possible spacetime paths (thus reflecting its "spread-out," wavelike nature).[20] While PTI considers the basic ontological quantum entity to be described by the state vector $|\Psi\rangle$ rather than by a wave function (which is a projection of the state vector onto the position basis), we can gain insight into the relationship of transactions to spacetime trajectories by considering Feynman's approach.

The Feynman sum-over-paths method asks the question: What are all the possible paths that a hypothetical particle could take from point A to a final point B? (We can think of A and B as spacetime points in a heuristic sense, but we should not assume that the particle "really" takes all paths as trajectories in a preexisting spacetime substance – remember that these are just *possible* paths, in the submerged-part-of-the-iceberg sense.) One then adds up all the possible paths in a particular way (reflecting that they have both magnitude and phase), giving what can be called the "Feynman amplitude" for getting from A to B. If there are no obstacles (i.e., absorbers) of any kind between A and B, it turns out that the path predicted by this procedure is the ordinary classical path between A and B – that is, the path that a baseball would follow. This path can be considered a classical trajectory because there is virtually no uncertainty about it: one can predict with an extremely high degree of precision where the object will be at any given time as it propagates from A to B. In fact, this "sum-over-paths" process is an application of the "principle of least action" (PLA), also sometimes known as Hamilton's principle, after William Hamilton, who formulated it. It says that nature chooses the path between two end points A and B for which the action (a quantity related to the difference between an object's kinetic and potential energies) is a minimum. (It turns out that such universal laws as Newton's laws of motion and laws of electromagnetism are derivable from this principle, so it is very powerful.)

The situation becomes more complicated (and interesting) when there are obstacles present, such as in the two-slit experiment discussed in Chapter 1. (The reader is referred to Feynman and Hibbs, 1965, section 1-4, for a detailed discussion of the path integral in the presence of various obstacles.[21]) The type of

---

[20] For an eminently readable and delightful introduction to this formulation, the reader is encouraged to consult Feynman's popular book *QED: The Strange Theory of Light and Matter* (1985).

[21] Feynman makes an interesting comment in section 1-3 of Feynman and Hibbs (1965) regarding his formulation of the calculation necessary to obtain the probability of an event. In distinguishing between observable and unobservable alternatives for a particle (where its path through one or the other slit falls into the "unobservable" category), he apparently wants to deny the following type of description: "When you watch, you find that it goes through either one or the other hole; but if you are not looking, you cannot say that it goes either one way

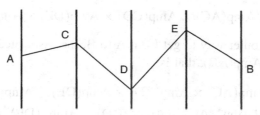

Figure 4.10 A path from A to B.

phenomenon that results depends on the nature of the system under study and the obstacles: specifically, whether its quantum wavelength (recall Chapter 1) is significant compared with the obstacles (and/or their separation). If that wavelength is significant, then we have a situation in which the single classical trajectory discussed above is replaced by several (or many) possible paths, with interference between them (i.e., there is no clearly defined trajectory).

Transactions can be considered simply a version of the Feynman sum-over-paths approach, with the added feature that absorption processes generate advanced confirmations which in turn give rise to weighted incipient transactions. Taking into account the confirmation at B requires us to multiply the Feynman (retarded, or future-directed) amplitude for A → B by the advanced amplitude for B → A, yielding the Born Rule. Just as in the Feynman amplitude for a quantum to get from A to B in the two-slit experiment, there is no well-defined spacetime trajectory. Such a trajectory exists only in an idealized classical (zero-wavelength) limit. While one can define the "amplitude of a path" for a quantum particle in the context of the Feynman picture, this does not correspond to a well-behaved probability unless there is a sequence of actualized transactions defining the associated trajectory. In what follows, we'll see why.

Using the Feynman sum-over-paths method, one obtains the probability to go from point A to B by summing over all possible "paths" from A to B to get an amplitude, and then squaring that amplitude. Let us simplify this as in the first chapter of Feynman and Hibbs (1965), wherein the space between A and B is subdivided by a finite number of intermediate stages, say, C, D, and E (Figure 4.10).

One first obtains the amplitude Amp(AC) to go from A to C, then similarly from C to D, from D to E, and finally from E to B. The total amplitude to go from A to B by way of C, D, and E is the product of these amplitudes:

or the other!" Yet his alternative description, in terms of formulated rules for calculating the probabilities, essentially boils down to the situation in the quoted sentence. Those rules simply substitute the presence or absence of a measuring apparatus for someone "looking" or "not looking."

$$\text{Amp(AB)} = \text{Amp(AC)} \times \text{Amp(CD)} \times \text{Amp(DE)} \times \text{Amp(EB)}. \qquad (4.2)$$

Now, if there is no other way to get from A to B, the associated probability is the absolute square of Amp(AB), that is,

$$\begin{aligned}
\text{Prob(AB)} = &\{\text{Amp(AC)} \times \text{Amp(CD)} \times \text{Amp(DE)} \times \text{Amp(EB)}\} \\
&\times \{\text{Amp}^*(\text{AC}) \times \text{Amp}^*(\text{CD}) \times \text{Amp}^*(\text{DE}) \times \text{Amp}^*(\text{EB})\}.
\end{aligned}$$
$$(4.3)$$

Mathematically, we can just rearrange this to get:

$$\begin{aligned}
\text{Prob(AB)} = &\{\text{Amp(AC)} \times \text{Amp}^*(\text{AC})\} \times \{\text{Amp(CD)} \times \text{Amp}^*(\text{CD})\} \\
&\times \text{Amp(DE)} \times \text{Amp}^*(\text{DE})\} \times \{\text{Amp(EB)} \times \text{Amp}^*(\text{EB})\}. \quad (4.4)
\end{aligned}$$

If there *are*, however, other ways to get from A to B, we still might be tempted to assume that we can define a "probability" to go between each of the intermediate stages, that is,

$$\text{Prob(AB)} = \text{Prob(AC)} \times \text{Prob(CD)} \times \text{Prob(DE)} \times \text{Prob(EB)}, \qquad (4.5)$$

implying that there exists a physically meaningful "probability for a particular path" such as A → C → D → E → B. However, under TI, the only reason you multiply an amplitude by its complex conjugate is because a confirmation occurs. If there is no absorber at the points C, D, or E, there is no independent complex conjugate factor such as Amp*(AC) and so on.

Moreover, such "partial amplitudes" as Amp(AC) do not correspond to well-behaved probabilities. This is well known in the context of the two-slit experiment, where the amplitudes to go from a source to a point x on a final screen by way of slits A or B do not correspond to probabilities that are additive (recall Section 3.3.2). In this sense, there is an "amplitude to go from the source to the slit by way of slit A (or slit B)," but that does not correspond to a meaningful probability that a particle actually went one way or the other, *unless* there are absorbers at the intermediate points (i.e., unless we have a detector to see "which slit the particle went through"). This *is* the case in a bubble chamber, so one can define a "path" for a quantum particle in a bubble chamber due to the interaction of the OW with absorbers in the chamber (we examine this case in the next section). We should not, however, let this lead us to think that outside the bubble chamber, the particle pursues a particular spacetime trajectory. The reason that you must add the amplitudes for all possible ways to go from A to B is because the quantum (i.e., offer wave) *does*, in a sense, pursue all those possible ways – it does this in the quantum substratum, not in spacetime.

4.4.2 *Trajectories in a Bubble Chamber*

A bubble chamber track is created by the ionization of molecules in the medium, which then act as catalysts for the formation of bubbles. What we actually see is a chain of bubbles forming around a chain of ionized molecules. In a typical bubble chamber interaction, a highly energetic quantum enters the chamber and is scattered by the first interacting molecule (in the process ionizing the molecule), but is not annihilated. Instead, it loses energy to the ionization process and continues on to subsequent molecules, repeating the process until all its energy has been "bled off."

Standard theoretical approaches to the passage of energetic charged particles through a medium use either the Bethe–Bloch equation (cf. Bethe, 1930) or the alternative Allison–Cobb (AC) approach (Allison and Cobb, 1980). The latter models the incoming particle as being surrounded by a cloud of virtual photons interacting with a dielectric medium (i.e., the atoms/molecules are polarized due to electromagnetic interactions). Classically, the photons are doing work against the field due to the polarized medium. Quantum mechanically, what is being calculated is the probability of energy transfer by photons of energy $E = \hbar\omega$, in each scattering of the incoming particle with a gas molecule.

The AC model lends itself to a transactional interpretation if we consider each scattering event as the exchange of OW and CW between the incoming particle and the target gas molecule. In effect, the incoming particle acts as an emitter of photon OW, and the gas molecule acts as an absorber (with the transferred energy being used to ionize the molecule). The probability of energy transfer for each scattering event (the square of the amplitude associated with each event) is simply the probability of a transaction; thus, the rate of energy loss is the rate of transactions transferring energy from the incoming particle to various gas molecules. The result is a chain of ionized molecules whose character reflects the specific properties of the incoming particle and the medium. The chain will be appropriately curved in the presence of a magnetic field, since the scattering computation takes into account whatever electromagnetic field is present in the medium. Thus we get the appearance of a "trajectory," which results from the ionization of gas molecules due to transfers of energy, via transactions, from the incoming quantum. But we should not let this mislead us into thinking that the incoming particle pursued a well-defined spacetime trajectory in the absence of the bubble chamber absorbers. I elaborate on this and related metaphysical points in Chapter 7.

4.5 Revisiting the Two-Slit Experiment

To conclude this chapter, I revisit the basic two-slit experiment in the context of the Feynman sum-over-paths picture. Here I consider only two possibilities: (1) no

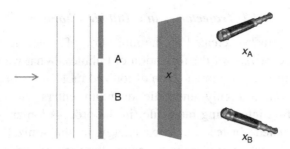

Figure 4.11 Two-slit experiment with an optional "which slit" measurement via telescopes.

"which slit" measurement at all or (2) a very sharp "which slit" measurement. However, as noted in the previous section, the quantum doesn't really pursue a "path"; results of measurements reflect transactions between an emitter and one or more absorbers, and do not imply or "reveal" a spacetime trajectory.

Returning to the basic two-slit setup (see Figure 4.11), the final detection screen can be considered as being composed of a large number of absorbers, each corresponding to all the possible positions x on the screen. Recalling the discussion in the previous section, we see that for large wavelengths and in the absence of a "which slit" measurement, the OW has a significant amplitude to interact with the absorbers defining the boundaries of both slits A and B, as does the CW generated by any of the absorbers in the final detection screen. That is, a CW is generated by each absorber x, and all such CW must be considered as having access to both slits (as opposed to only one or the other slit). The probability of detection at a particular position x is the product of the OW amplitude at x (which is the OW component reaching x as opposed to some other position x') and the CW amplitude generated at x and terminating (by way of both slits) at the emitter. In more quantitative terms, the amplitude of the OW reaching a position x can be represented as $\langle x|\Psi\rangle = \frac{1}{\sqrt{2}}[\langle x|A\rangle + \langle x|B\rangle]$ and the amplitude of the CW generated at x as $\langle\Psi|x\rangle = \frac{1}{\sqrt{2}}[\langle A|x\rangle + \langle B|x\rangle]$.

The probability given by the product of the amplitudes is therefore (where we include the projection operator resulting from the outer product of the brac and ket):

$$P(x)|x\rangle\langle x| = \langle x|\Psi\rangle\langle\Psi|x\rangle|x\rangle\langle x| = \frac{1}{2}[\langle x|A\rangle + \langle x|B\rangle][\langle A|x\rangle + \langle B|x\rangle]|x\rangle\langle x|$$

$$= [\langle x|A\rangle\langle A|x\rangle + \langle x|B\rangle\langle B|x\rangle + \langle x|A\rangle\langle B|x\rangle + \langle x|B\rangle\langle A|x\rangle]|x\rangle\langle x|$$

(4.6)

where the last two terms reflect interference between the slits.

Now suppose we consider a "which slit" measurement of the kind envisioned in Wheeler's famous "delayed choice" experiment. This consists of replacing the final detection screen with a pair of telescopes, each focused on one of the slits A and B. (This version is done with photons.) What the focusing mechanism does, in terms of the (time-reversed) Feynman paths picture, is to greatly increase the amplitude for a CW to interact with the absorbers defining the boundaries of the slit at which the telescope is aimed, while making the amplitude for interaction with the other slit negligible. This means that, even though the OW has a finite amplitude to interact with both slits, the CW generated by either telescope does not.[22] (One can also see this as the OW component corresponding to slit A having negligible amplitude to reach telescope B and vice versa.) Again in more quantitative terms, the probability of detection at a particular telescope x_A must be specified in the absorber basis and is given by

$$P(x_A)|x_A\rangle\langle x_A| = \langle x_A|\Psi\rangle\langle\Psi|x_A\rangle|x_A\rangle\langle x_A|$$
$$= \frac{1}{2}[\langle x_A|A\rangle\langle A|x_A\rangle + \langle x_A|B\rangle\langle B|x_A\rangle]|x_A\rangle\langle x_A| \qquad (4.7)$$

which exhibits no interference, since $\langle x_A|B\rangle$ is zero. The point is that these probabilities are reflections of physical amplitudes of interactions between emitters and absorbers, and do not indicate spacetime trajectories. No well-defined particle trajectory can be inferred based on amplitudes that apply to the pre-spacetime level. For example, the OW does have a finite amplitude to interact with either slit boundary (i.e., both $\langle\Psi|A\rangle$ and $\langle\Psi|B\rangle$ are different from zero), so one can think of the OW as "having gone through" both slits, but since the CWs do not (i.e., $\langle B|x_A\rangle = \langle A|x_B\rangle = 0$), the transactions available do not exhibit two-slit interference. The "particle" is no more and no less than whatever transaction is actualized.

4.6 Null Measurements

Finally, let us consider null measurements. These are cases in which the system of interest fails to be detected at a particular detector, from which we can infer that it "went the other way." Such situations appear paradoxical in the standard approach to quantum theory, as we'll review below. However, when we take into account that quantum systems are really *precursors* to spacetime events and that such events emerge through the transactional process, null measurements are perfectly natural aspects of garden-variety measurements.

Perhaps the earliest example of a null measurement was proposed by Mauritius Renninger in 1960. He imagined a photon source surrounded by two concentric

---

[22] Technically, the "absorber" for each telescope is a macroscopic object comprising many microscopic absorbers. Here the phrase "telescope at x_A" just means the entire class of microscopic absorbers corresponding to telescope A.

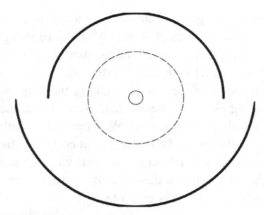

Figure 4.12 The Renninger experiment.

hemispheres, both functioning as detectors. One of these is closer to the source than the other, as shown in Figure 4.12.

In the usual approach, the photon is described only by a spherically expanding wave function that (somehow) collapses upon detection of the photon. If the photon is not detected at the inner hemisphere, that means it is certain to be detected at the outer one. But (so the usual story goes) this means that the photon's wave function partially collapsed as a result of the apparent lack of a measurement – that is, its nondetection at the inner hemisphere. How could this happen?

In the transactional picture the paradox vanishes, since there *is* interaction between the source and both hemispheres, at the level of possibility. All absorbers in both hemispheres respond with confirmations to the component of the photon's offer wave that interacts with them (see Chapter 5 for quantitative specifics on how these components are defined with respect to each responding absorber). This gives rise to a set of incipient transactions, half of which involve the inner hemisphere and half the outer hemisphere. But again, these are pre-spacetime processes; nothing happens in spacetime until one of the incipient transactions is actualized.[23] Once one of these is actualized with the appropriate probability, the photon proceeds as a quantum of real spacetime-situated energy from the emitter to the "winning" absorber and is detected there. Thus, there definitely has been a complete measurement that involved interactions with both hemispheres. The null

[23] Philosophers might wonder at the use of seemingly temporal language here in the use of "until." This should be understood in the sense of a Whiteheadian process. Processes can occur without necessarily being spacetime processes to which one can attribute the usual metrical time and localization indices. This point is emphasized by J. Seibt (2020), who notes: "if spacetime is quantized and emergent, metaphysics cannot operate with basic entities that are individuated in terms of their spacetime locations."

measurement simply reflects that the transaction that was actualized involved an absorber in the outer hemisphere.

The conceptual challenge of this account is our usual preconception that physical interactions are necessarily spacetime processes and/or that the photon is always a spacetime-situated object, either a particle or a wave heading unidirectionally from the source outward. The second preconception precludes any active participation by the receiving components; that is, the absorbers. However, quantum theory requires that we renounce those classical preconceptions. Instead, we need to think of the spacetime *theater* as just that – the final phenomenal result of a plethora of unseen activity going on "behind the scenes" to create the scene that we can experience as some aspect of the classical world of everyday macroscopic phenomena. The behind-the-scenes quantum processes are described by complex numbers and multidimensional entangled states that, in a strict mathematical sense, are simply too big to fit on the stage of 3+1 spacetime. Only the final actualized measurement result is the spacetime-situated process.

In the next chapter, we deal with specifics of the relativistic expansion of TI, which has become known in the literature as "RTI." Both terms, PTI and RTI, refer to the same model. They simply emphasize different aspects.

5

The Relativistic Transactional Interpretation

5.1 TI Has Basic Compatibility with Relativity

As noted in Cramer (1986), the original version of TI already has basic compatibility with relativity in virtue of the fact that the actualization of a transaction occurs with respect to the end points of a spacetime interval or intervals, rather than at a particular instant of time, the latter being a non-covariant notion. Its compatibility with relativity is also evident in that it makes use of both the retarded and advanced solutions obtained from the Schrödinger equation and the complex conjugate Schrödinger equation, respectively, both of which are obtained from the relativistic Klein–Gordon equation by alternative limiting procedures. Cramer (1980, 1986) has noted that, in addition to Wheeler and Feynman, several authors (including Dirac) have laid groundwork for and/or explored explicitly time-symmetric formulations of relativistic quantum theory with far more success than has generally been appreciated.[1] This chapter is largely devoted to developing the transactional picture in terms of a quantum relativistic extension of the Wheeler–Feynman theory by Davies (1970, 1971, 1972). It is somewhat more technical than the previous chapters.

The crucial feature of TI that allows it to "cut the Gordian knot" of the measurement problem is that it includes the role of absorbers as active participants in a real physical process that must be included in the theoretical formalism in order to account for the process of measurement. The preceding is a specifically relativistic aspect of quantum theory, since nonrelativistic quantum mechanics (NRQM) ignores absorption: it addresses only persistent particles. Strictly speaking, NRQM ignores emission as well; there is no formal component of the nonrelativistic theory corresponding to an emission process. The theory is applied only to an entity or entities assumed to be already in existence. In contrast, relativistic quantum field theory explicitly includes emission and absorption

[1] For example, Dirac (1938), Hoyle and Narlikar (1969), Pegg (1975), Konopinski (1980), Bennett (1987).

through the field creation and annihilation operators, respectively; there are no such operators in nonrelativistic quantum mechanics. Because NRQM treats only preexisting particles, the actual emission event is not included in the theory, which simply applies the ket $|\Psi\rangle$ to the preexisting system under consideration. Under these restricted circumstances, it is hard to see a physical referent to the brac $\langle\Psi|$ from within the theory, even though it enters computations needed to establish empirical correspondence. What TI does is to "widen the scope" of nonrelativistic quantum theory to take into account both emission and absorption events, the latter giving rise to the advanced state or brac $\langle\Psi|$. In this respect, again, it is harmonious with relativistic quantum theory.

It should also be noted that the standard notion of field generation as being isotropic with respect to space (i.e., a spherical wave front) but *not* isotropic with respect to time (i.e., that emission is only in the *forward* light cone) seems inconsistent and intrinsically ill-suited to a relativistic picture, in which space and time enter on an equal footing (except, of course, for the metrical sign difference). The prescription of the time-symmetric theory for half the generated field in the $+t$ direction and half in the $-t$ direction is consistent with the known fact that (in general) field generation does not favor one spatial direction over another, and harmonious with the relativistic principle that a spacetime point is a unified concept represented by the four-vector $x^\mu = \{x^0, x^1, x^2, x^3\}$.

5.2 The Quantum Direct-Action Theory

We first consider the theory of Davies, which provides a natural starting point for TI in the relativistic domain. In what follows, we shall refer to the relativistic generalization of TI as the "Relativistic Transactional Interpretation" (RTI). The terms "PTI" and "RTI" can be used interchangeably to refer to the same interpretation proposed herein. They simply emphasize different aspects of the formulation.

5.2.1 Preliminary Remarks

The Davies theory has been termed an "action at a distance" or "direct-action" theory because, like the classical Wheeler–Feynman theory, it describes interactions not in terms of a mediating field with independent degrees of freedom, but rather in terms of direct interactions between currents (field sources). One of the original motivations for such an "action at a distance" theory was to eliminate divergences, stemming from self-action of the field, from the standard theory. However, it was later realized that some form of self-action was needed in order to account for such quantum phenomena as the Lamb shift. The Davies theory does allow for self-action in that a current can be regarded as acting on itself in the case of indistinguishable currents (see, e.g., Davies, 1971, p. 841, figure 2).

Nevertheless, despite its natural affinity for a time-symmetric model of the field, it must be emphasized that RTI does *not* involve an ontological elimination of the field. The field concept remains as a measure of the direct interaction between charges, quantifiable in terms of a potential. The basic field is nonquantized, but quantization emerges under suitable conditions. We return to this issue below.

Thus, RTI is based not on elimination of fields, but rather on the time-symmetric, transactional character of energy propagation by way of those fields, and the feature that offer and confirmation waves capable of resulting in empirically detectable transfers of physical quantities only occur in couplings between field currents. However, in keeping with this possibilist reinterpretation, the fields themselves are considered as pre-spacetime objects. That is, they exist, but not as spacetime entities. Rather, spacetime entities are restricted to actualized, detectable conserved currents: real-valued energy/momentum transfers. At first glance, this ontology may seem strange; however, when one recalls that such standard objects of quantum field theory as the vacuum state $|0\rangle$ have no spacetime arguments and are maximally nonlocal,[2] it seems reasonable to suppose that such objects exist, but not in spacetime (in the sense that they cannot be associated with any region in spacetime).

Shimony has similarly suggested that spacetime can be considered as a domain of actuality emergent from a quantum level of possibilities:

There may indeed be "peaceful coexistence" between Quantum nonlocality and Relativistic locality, but it may have less to do with signaling than with the ontology of the quantum state. Heisenberg's view of the mode of reality of the quantum state was ... that it is *potentiality* as contrasted with *actuality*. This distinction is successful in making a number of features of quantum mechanics intuitively plausible – indefiniteness of properties, complementarity, indeterminacy of measurement outcomes, and objective probability. But now something can be added, at least as a conjecture: that the domain governed by relativistic locality is the domain of actuality, while potentialities have *careers* in space-time (if that word is appropriate) which modify and even violate the restrictions that space-time structure imposes upon actual events. *(2009, section 7, item 2)*

Shimony goes on to note the challenges in providing an account of the emergence of actuality from potentiality, which amounts to "collapse" or quantum state reduction. RTI suggests that transactions are the vehicle for this process[3] and that

---

[2] This is demonstrated by the Reeh–Schlieder Theorem; cf. Redhead (1995).

[3] Recall that even if no specific "mechanism" is provided for the actualization of a transaction, TI provides a solution to the measurement problem in that it ends the usual infinite regress by taking into account absorption, which physically determines the measurement basis. A measurement is completed when absorption occurs, and the conditions for that can be precisely specified, as shown in this chapter. Moreover, as suggested above, it is

aspects of it must involve processes and entities transcending the spacetime construct. We can call this domain of possibilities the *quantum substratum*.

A further comment is in order regarding the proposal that spacetime is emergent rather than fundamental. In the introductory chapter to their classic *Quantum Electrodynamics*, Beretstetskii, Lifschitz, and Petaevskii make the following observation concerning QED interactions:

> For photons, the ultra-relativistic case always applies, and the expression $[\Delta q \sim \hbar/p]$, where Δq is the uncertainty in position, is therefore valid. This means that the coordinates of a photon are meaningful only in cases where the characteristic dimension of the problem is large in comparison with the wavelength. This is just the "classical" limit, corresponding to geometric optics, in which the radiation can be said to be propagated along definite paths or rays. In the quantum case, however, where the wavelength cannot be regarded as small, the concept of coordinates of the photon has no meaning....
>
> The foregoing discussion suggests that the theory will not consider the time dependence of particle interaction processes. It will show that in these processes there are no characteristics precisely definable (even within the usual limitations of quantum mechanics); *the description of such a process as occurring in the course of time is therefore just as unreal as the classical paths are in non-relativistic quantum mechanics.* The only observable quantities are the properties (momenta, polarization) of free particles: the initial particles which come into interaction, and the final properties which result from the process. [The authors then reference L. D. Landau and R. E. Peierls, 1930.[4]] *(1971, p. 3, emphasis added)*

The italicized sentence asserts that the interactions described by QED (and, by extension, by other interacting field theories) cannot consistently be considered as taking place in spacetime. Yet they do take place *somewhere*; the computational procedures deal with entities implicitly taken as ontologically real. This "somewhere" is just the pre-spatiotemporal, pre-empirical realm of possibilities – the quantum substratum – proposed herein.

5.2.2 Background

In this section, we will first review the basic classical absorber or "direct-action" theory and a semi-classical quantum version due to Davies (1971, 1972). It should be noted that Davies' treatment, while an advance in the quantum direction from the original classical Wheeler–Feynman theory, remained semi-classical insofar as

---

likely misguided to demand a causal, mechanistic account of collapse, since as Shimony suggests, one is dealing with a domain that transcends the causal spacetime realm. We depart slightly from Shimony's formulation just in noting that quantum entities don't have "careers" in spacetime but rather exist in the quantum substratum.

[4] The Landau and Peierls paper has been reprinted in Wheeler and Zurek (1983).

it tacitly identified radiation with continuous fields. It also assumed that a real photon could be unilaterally emitted, which, as we shall see, is not the case at the quantum level. Thus, ambiguity remained in that account regarding the distinction between real and virtual photons, as well as the nature of the relevant absorber boundary condition, or so-called light-tight box condition, which has led to some confusion. However, it is a useful starting point for the present work, which revises certain features pertaining to the quantization of the radiated field. The revised account makes clear the fully quantum nature of the appropriate boundary condition, which is really a particular sort of emitter/absorber interaction rather than any specific configuration of absorbers as is implied by the usual term "light-tight box."

We first review standard classical electromagnetic theory. The standard way of representing the field A acting on an accelerating charge i due to other charges j is as the sum of the retarded fields due to j and a "free field":

$$A = \sum_{j \neq i} A^{ret}_{(j)} + \frac{1}{2}\left(A^{ret}_{(i)} - A^{adv}_{(i)} \right). \tag{5.1}$$

In the classical expression (5.1), self-action is omitted to avoid infinities (which are dealt with in quantum field theory by renormalization). $A^{ret}_{(j)}$ is the retarded solution to the inhomogeneous equation, that is, the field equation with a source, while the second term pertaining only to i is a solution to the homogeneous field equation (source-free). The latter quantity, the "radiation term," is originally due to Dirac and is necessary in order to account for the loss of energy by a radiating charge if it is assumed that all sourced fields are retarded only. Wheeler and Feynman (1945) critically remark in this regard:

The physical origin of Dirac's radiation field is nevertheless not clear. (a) This field is defined for times before as well as after the moment of acceleration of the particle. (b) The field has no singularity at the position of the particle and by Maxwell's equations must, therefore, be attributed either to sources other than the charge itself or to radiation coming in from an infinite distance. (*p. 159*)

These authors' concern about the source of Dirac's radiation field is resolved in the direct-action theory (DAT). The classical direct-action or "absorber" theory proposed that the total field $A^{(DA)}$ acting on i is given by:

$$A^{(DA)} = \sum_{j \neq i} \frac{1}{2}\left(A^{ret}_{(j)} + A^{adv}_{(j)} \right); \tag{5.2}$$

that is, it is given by the sum of the time-symmetric fields generated by all charges *except* i. Absorbing charges respond to the emitted field with their own time-symmetric field, contributing to the sum in (5.2). Wheeler and Feynman noted that (5.1) and (5.2) are equivalent provided that their difference is zero, that is,

$$\sum_{\forall j} \frac{1}{2} \left(A_{(j)}^{ret} - A_{(j)}^{adv} \right) = 0. \tag{5.3}$$

Under the condition (5.3), the responses of absorbing charges to the time-symmetric field of the emitting charge yields an effective "free field" applying only to the emitting charge, that is, the second term of (5.1). It's important to note that this term attributes a solution to the homogeneous equation to a particular charge that is (of course) not its source, as observed by WF above. In the DAT, the "free field" is actually sourced by other charges (responding absorbers) and only *appears* to have the form of a free field from the standpoint of the accelerating charge whose index it bears.

The condition (5.3) is historically termed the "light-tight box" condition (LTB) in the classical theory. It is commonly interpreted as the constraint that "all radiation is absorbed," but this characterization is misleading even at the classical level and requires explicit reformulation at the quantum level. For one thing, it conflates the static, time-symmetric Coulomb field with a dynamic radiation field.[5] In addition, the mathematical content of (5.3) says only that the net radiation field is zero. This can just as legitimately be interpreted to mean that *there is no true free (unsourced) radiation field*. While selective cancellation of fields does occur among charges to produce the effective local radiation field, the absence of an unsourced radiation field is the primary physical content of the LTB condition for the quantum form of the DAT, as we will see in Section 5.2.4.

Other weaknesses in the original classical DAT have been discussed by Gründler (2015), who notes that field cancellation via explicit evaluation of the interactions between the emitter and the other charges depends on imposing an arguably unjustified asymmetrical condition: an effective index of refraction applying only to absorber responses. He argues that the equivalence between the classical DAT and standard classical electrodynamics for individual charges amounts only to a formal one based on (5.2) and (5.3).

In any case, the ambiguity inherent in the classical treatment, and the practice of interpreting (5.3) as being about some specific distribution of charges, has led to some confusion regarding the nature of the relevant condition – the analog of (5.3) – pertaining to the quantum case. In what follows, we clarify the nature of the quantum version of the direct action theory (QDAT) and define the applicable boundary conditions.

---

[5] Actually, the classical DAT appears to assume that even the time-symmetric fields are present only in the case of an accelerating charge, which neglects the static Coulomb interaction.

5.2.3 The Quantum Direct-Action Theory: Basics

In this section, we will discuss the DAT in terms of Green's functions or "propagators" (solutions to the field equation for a point source, and related source-free forms), since that is the natural way to formulate the QDAT. It should be noted that, in contrast to the field $A(x)$ with a single argument, propagators are functions of two arguments and always relate two specific coordinate points. In standard quantum field theory, propagators are correlation functions for pairs of field coordinates.[6] In the QDAT, propagator arguments are parameters of the source currents (charges).

The corresponding quantities are:

$D_{ret}(x - y)$: retarded solution to the inhomogeneous equation
$D_{adv}(x - y)$: advanced solution to the inhomogeneous equation
$\bar{D}(x - y) = \frac{1}{2}(D_{ret} + D_{adv})$: time-symmetric solution to the inhomogeneous equation
$D(x - y) = (D_{ret} - D_{adv})$: odd solution to the homogeneous equation

In terms of these, we can see that the following identity holds:

$$D_{ret} = \bar{D} + \frac{1}{2}D. \tag{5.4}$$

This describes the elementary field of a single charge in the DAT, taking into account the "response of the absorber" corresponding to the second term. It differs from (5.1) in that it does not exclude the charge from the effects of the field. As noted by Wheeler and Feynman (1945), the first term is singular, which is why they sought to prohibit a charge from interacting with its own field. At the quantum level, in view of indistinguishability, one cannot impose such a restriction, since in general one cannot define which charge is the source of the time-symmetric field.[7] So the self-action involving the time-symmetric field must be retained at the quantum level (however, we will see later that this self-action does not involve any exchange of real energy and consists only of self-force). This expression shows how a net retarded field arises due to the combination of absorber response (an effective "free field" acting on the emitting charge) with the basic time-symmetric field of the emitting charge. We now investigate the analogous situation in the QDAT.

---

[6] As suggested by Auyang (1995), these coordinates are best understood as parameters of the field, rather than "locations in spacetime." The same understanding can be applied to the nonquantized field of the QDAT, in which field sources are the referent for the parameters.

[7] Charges become distinguishable only in situations in which energy–momentum conservation is satisfied and non-unitarity can occur. This issue is elaborated in Section 5.2.5.

First, it is important to note that the propagators defined above make no distinction between positive and negative frequencies, since the classical theory assumes that all frequencies are positive and makes no connection between frequency and energy. However, the quantum theory of fields must explicitly deal with the existence of positive and negative frequencies. Thus, in the QDAT, each of the quantities above must be understood as comprising positive- and negative-frequency components. Since there are many different conventions for defining these quantities, we write the components here explicitly in terms of vacuum expectation values or "cut propagators" $\Delta^{\pm}$. In these terms,

$$iD(x - y) \equiv i\Delta(x - y) = \langle 0|[A(x), A(y)]|0\rangle$$
$$= (\langle 0|A(x)A(y)|0\rangle - \langle 0|A(y)A(x)|0\rangle) \equiv (\Delta^+ - \Delta^-) \qquad (5.5)$$

where $A(x)$ is the usual quantum electromagnetic field, and under Davies' convention for the components, we define

$$D(x - y) = D^+ + D^- = (-i\Delta^+) + (i\Delta^-). \qquad (5.6)$$

Note in particular, for later purposes, that D^- is defined with the opposite sign of the negative-frequency cut propagator Δ^-:

$$iD^-(x - y) \equiv -\Delta^-(x - y) = -\langle 0|A(y)A(x)|0\rangle. \qquad (5.7)$$

We also need the even solution to the homogeneous equation, D_1 (cf. Bjorken and Drell, 1965, appendix C):

$$D_1(x - y) = i(D^+(x - y) - D^-(x - y)) = \Delta^+(x - y) + \Delta^-(x - y). \qquad (5.8)$$

Note that each of the positive- and negative-frequency components of these fields independently reflects the same relationship of retarded and advanced solutions as the total field, for example, $D^+(x - y) = (D^+_{ret} - D^+_{adv})$.

Feynman's innovation was to interpret negative frequencies as antiparticles, specifically, as "particles with negative energies propagating into the past." This is equivalent to antiparticles with positive energy propagating into the future, where antiparticles have the opposite charge (cf. Kastner, 2016a). To that end, he defined a propagator that does just that, that is, assigns the retarded propagator only to positive frequencies and the advanced propagator only to negative frequencies. The result is the "Feynman propagator," D_F:

$$D_F = D^+_{ret} + D^-_{adv}. \qquad (5.9)$$

This satisfies an identity analogous to (5.4):

$$D_F = \bar{D} - \frac{i}{2}D_1. \qquad (5.10)$$

To see (5.10) explicitly, we write the quantities in terms of their positive- and negative-frequency components, using (5.8) for D_1:

$$\bar{D} - \frac{i}{2}D_1 = \frac{1}{2}\left[(D_{ret}^+ + D_{adv}^+) + (D_{ret}^- + D_{adv}^-)\right] + \frac{1}{2}\left[(D_{ret}^+ - D_{adv}^+) - (D_{ret}^- - D_{adv}^-)\right]$$
$$= (D_{ret}^+ + D_{adv}^-) = D_F. \tag{5.11}$$

5.2.4 "Light-Tight Box" Condition Modified at Quantum Level

As observed by Davies (1971), a basic quantum version of the direct-action theory (QDAT) has actually been around since Feynman (1950). Feynman showed that for the case when the number n of external (commonly termed "real") photon states is zero, the standard quantum action J for the interaction of the quantized field $\hat{A}$ with a current j can be replaced by a direct current-to-current interaction, as follows:

$$J(n = 0) = \sum_i \int j_{(i)}^\mu(x)\hat{A}_\mu(x)\, d^4x \quad = \sum_i \sum_j \frac{1}{2} \int j_{(i)}^\mu(x)D_F(x - y)j_{\mu(j)}(y)\, d^4x\, d^4y$$

$$\text{(5.12a, b)}$$

where D_F is the Feynman propagator as defined in (5.9) and (5.10). Davies notes that the same result is proved by way of the S-matrix in Akhiezer and Berestetskii (1965, p. 302). So it is important to note that (5.12) *is a theorem* and holds even if one has started from the usual assumption that there exists an independent quantum electromagnetic field $\hat{A}$.

Now, the entire content of the so-called light-tight box condition (LTB) for the quantum version of the direct action theory (QDAT) is contained in the condition for the replacement of (a) by (b) in (5.12). But the LTB condition has traditionally been deeply mired in ambiguity about what sort of entity counts as a "real photon" and about what physical situations give rise to real photons. It has additionally been hampered by a semi-classical notion of "absorption of radiation." However, it is straightforward from the mathematics that what is actually required for the required replacement is simply the nonexistence of an independent quantized electromagnetic operator field $\hat{A}$ – that is, vanishing of the usual postulated system of oscillators of standard quantum field theory! We can see that explicitly by way of the proof of Akhiezer and Berestetskii, who obtain an expression for the scattering matrix S in the general case, with no restriction. That expression is:

$$S = \exp\left(-\frac{i}{2}\int j^\mu(x)D_F(x - y)j_\mu(y)\, d^4x\, d^4y\right) \times \exp\left(i\int j_\mu(x)\hat{A}^\mu(x)\, d^4x\right)$$

$$\text{(5.13)}$$

where the usual chronological ordering of quantum field operators is understood, and $\hat{A}^\mu$ is the usual quantized electromagnetic field. Akhiezer and Berestetskii then say: "In processes in which no photons participate, the last factor is equal to unity, and the scattering matrix assumes the form [first factor only, as in Equation (5.12b)]." But again, this brings in the ill-defined notion of "participation of photons," when what is really done to obtain the final result is to simply *set the independent quantized electromagnetic field $\hat{A}^\mu$ to zero*. The crucial point, then, is the following: essentially all there is to the so-called light-tight box condition for the QDAT expressed in terms of the Feynman propagator D_F is *Wheeler and Feynman's original proposal to eliminate the electromagnetic field as an independent mechanical system*. Note that this corresponds to the condition (5.3) as interpreted in the previous section, that is, that there simply are no genuinely unsourced "free fields." Rather, any effective field of the form D (or D_1 for the QDAT) is obtained through a specific kind of interaction between sources, that is, between emitters and absorbers.[8]

In the next section we examine the QDAT in more detail, resolving some ambiguities about the distinction between real and virtual photons and discussing the relevance of the distinction for the quantum form of the LTB condition. We'll see that the only additional condition for equivalence of the QDAT with the standard theory amounts to the quantum completeness condition (and an appropriate phasing of the fields of the emitter and absorbers), which assures recovery of the Feynman propagator D_F.

5.2.5 Relativistic Generalization of Absorber Response

The Feynman propagator D_F is the quantum analog of (5.1); it reflects a "causal" field directed from smaller to greater temporal values for the case of positive frequencies and the opposite – from greater to smaller temporal values – for negative frequencies, with an effective "free field" for radiative processes. D_F arises due to the quantum relativistic analog of absorber response, which differs from the classical theory in two crucial respects. One is the need to take into account negative frequencies not present in the classical case, which requires separate phasing of the positive- and negative-frequency field components and leads to D_1 rather than D for the free field, as discussed above. The other is the mutual, or dynamically symmetric, emitter/absorber interaction giving rise to the "free field" D_1. To clarify the second point: at the relativistic level (which is the level at which nature really operates), emitters and absorbers participate

---

[8] Several authors have noted that one need not view the "zero-point energy" as evidence for an independently existing field, since exactly the same effects attributed to zero-point energy arise in the QDAT. See, for example, Bennett (1987) and Jaynes (1990).

together in the generation of offers and confirmations. Offer waves (OW) are not emitted unilaterally and *then* responded to; instead, *both* OW and CW are generated in a more symmetrical, mutual interaction that is non-unitary.[9] Importantly, this non-unitary interaction – the generation of offers and confirmations giving rise to a real, on-shell photon described by D_1 – must be carefully distinguished from the basic time-symmetric field connection $\bar{D}$, which is unitary. The latter corresponds *only* to an off-shell, virtual photon, that is, to the Coulomb field (zeroth component of the electromagnetic potential).

For clarity in the discussion regarding which process is under consideration, let us use the term "U-interaction" to denote the unitary, Coulomb, virtual photon interaction described by $\bar{D}$, and the term "NU-interaction" to denote the non-unitary, real photon interaction described by D_1. The latter is the relativistic analog of the "absorber response" of the nonrelativistic level. The former, basic U-interaction obtains in situations that do not satisfy energy conservation, for example, between two free electrons that would not be able to transfer real energy between them. The U-interaction conveys only force, not energy.

In contrast, a NU-interaction corresponds to radiative processes only, that is, to transversely polarized "free" fields or real photons describable by Fock states (more precisely, projection operators; this point is elaborated below). The latter type of process occurs with a well-defined probability – basically a decay rate. It occurs only when energy-momentum (and angular momentum) conservation is satisfied. Under these conditions, participating charges attain distinguishability, in that one is clearly losing conserved quantities and others are (possibly) gaining them. Here, we must say "possibly" because many absorbers are responding with CW, but in the case of a single photon, only one absorber can actually gain the conserved quantities transferred. (This issue, involving probabilistic behavior, is elaborated further below.)

Thus, in the QDAT, whenever there is only a static Coulomb field, it means that non-unitarity has *not* occurred; this is a U-interaction. The virtual photons that mediate the Coulomb field are not Fock states and thus are not described by offer or confirmation waves. For virtual photons, there is no fact of the matter about which current is emitting and which current is absorbing, since there is no OW or CW at this level. It is a force-only interaction and is not radiative. A useful mnemonic for this distinction is "virtual photons convey only force, while real photons convey energy." The fact that the unitary, time-symmetric connection

---

[9] As gauge bosons, real (on-shell) photons are only created through the non-unitary process described herein. Fermionic sources can be on-shell in the absence of a non-unitary process; thus, one can have a real electron without an "electron CW." This important distinction between bosons and their fermionic sources is elaborated in Section 6.1.

conveys only force explains why any divergences associated with the self-interaction do not involve real energy; they are force-divergences only.

In introducing this concept of the generation of a real photon – the NU-interaction – we come to an important previously "missing link" in extending the transactional picture to the relativistic level. This observation addresses and resolves a common criticism that emitters and absorbers are "primitive" and that absorber response is just a placeholder for the "external observer" in the measurement problem. On the contrary, the behavior of emitters and absorbers that trigger the non-unitary measurement transition is not "external" to the theory at all. It is fully accounted for and quantified within the relativistic QDAT in terms of the coupling amplitude or charge e. This issue is discussed in detail in Kastner (2018) and in Kastner and Cramer (2018). For current purposes, we note that the charge e is the basic amplitude for a photon to be emitted or absorbed, as previously observed by Feynman (1985). In the context of RTI and the QDAT, the interpretation is just slightly different: the coupling amplitude or charge is the *amplitude for either an OW or CW to be generated*. Since, as described above, one needs *both* the OW and CW to create a real (on-shell) photon in the QDAT, one has two factors of the charge e. Hence the basic probability of real photon generation – the NU-interaction – is the fine structure constant $\alpha = e^2$.

The foregoing highlights the crucial physical role of the fine structure constant in governing radiative processes. As noted above, when the NU-interaction does not occur, one still has the basic time-symmetric connection $\bar{D}$ corresponding to a virtual (off-shell) photon mediating to the static Coulomb field. Thus, a field can certainly be generated as the basic connection $\bar{D}$ between currents, but with no radiation (no real photon and no real energy). The crucial point: *field generation in the QDAT is not necessarily radiated energy*. Radiation is emitted *only* if the NU-interaction occurs, and it need not and often does not occur. The basic probability of its occurrence is the fine structure constant. For a specific situation, the relevant transition amplitudes from initial to final states of the emitter and absorber provide an additional factor, ultimately leading to a decay rate (see Kastner and Cramer, 2018).

Another important distinction between the classical DAT and the QDAT is that the relevant quantity for describing the interaction is the scattering matrix $S = Pe^{-iJ}$ (where J is the action and P a time-ordering operator). S defines probability amplitudes for transitions between initial and final states. This probabilistic behavior does not exist at all in the classical DAT, but is a crucial aspect of the QDAT. Its importance in differentiating the quantum situation from the original WF theory cannot be overstated. Failure to appreciate the entry of quantum probabilities into the field behavior leads to much confusion regarding what is meant by terms like "emission" and "radiation" and can lead to serious misunderstandings. In particular, in the quantum case one must distinguish between (1) the generation of a field, which could be just the static Coulomb field mediated by virtual photons through the bound field $\bar{D}$ (the U-interaction), and (2) the actual emission or radiation of a real (transversely polarized) photon, which occurs only for the NU-interaction giving rise to the "free field" D_1. In the classical case, there is no distinction between generating a field and radiating, since it is assumed that absorber response *always* occurs and that all generated fields are radiative in nature – that is, they convey electromagnetic energy corresponding to the intensity of the field. However, this is *not* the case at the quantum level, since (as noted above) the $\bar{D}$ field can exist as a basic connection between currents without any corresponding radiation or energy transfer.

The need for a probabilistic description arises because in the quantum case, one must take into account that the field itself is not equivalent to a "photon" in that a photon is discrete while the field is continuous (at least with respect to the parameter x). As an illustration, suppose we are dealing with a field state corresponding to one photon. Such a field in general propagates between an emitter and many absorbers; many absorbers can respond, even though there is only one photon "in the field." While the responses contribute to the creation of the real photon through the NU-interaction, the photon itself cannot go to all the responding absorbers; only one can actually receive it. This is where the probabilistic behavior, described by $S = Pe^{-iJ}$, enters. We make this issue more quantitative in what follows.

Looking at the Fourier components, one again sees that the Feynman propagator is complex, with both real and imaginary parts:

$$D_F(x) = \frac{1}{(2\pi)^4} \int \left(\frac{PP}{k^2} - i\pi\delta(k^2) \right) e^{ikx} dk = \bar{D}(x) - \frac{i}{2}D_1(x) \qquad (5.14)$$

(*PP* stands for the principal part). The complexity of D_F implies intrinsic non-unitarity, a point whose implications we will consider in Section 5.2.6. The real part $\bar{D}$ is the time-symmetric propagator, while the imaginary part D_1 is the even "free field" or solution to the homogeneous equation as defined above.[10]

As Davies notes, "The $\bar{D}$ part (bound field) leads to the real principal part term which describes virtual photons ($k^2 \neq 0$), whilst the imaginary part D_1 (free field) describes photons with $k^2 = 0$, that is, real photons, through the delta function term." (Davies, 1972, p. 1027). The D_1 term is the quantum analog of the free field in Equations (5.1) and (5.4). In the classical DAT, the "free field" is assumed to be present for all accelerated particles due to the "response of the universe" or "absorber response." In order to understand the circumstances and physical

---

[10] Here, we are using the sign conventions in Bjorken and Drell (1965, appendix C).

meaning of the D_1 interaction for the QDAT, we must clearly define the quantum analog of acceleration and distinguish that from the static case, in which only the Coulomb (nonradiative) interaction $\bar{D}$ is present. The quantum analog of acceleration is a state transition, such as from a higher to a lower atomic energy state, accompanied by radiation. In contrast, for the static case, there is no radiation, so there is no effective free field – no "absorber response." Thus, in the QDAT, the presence or absence of "absorber response" – really the mutual NU-interaction, as discussed above – is what dictates whether there will be a D_1 component and hence a quantum form of acceleration accompanied by radiation (i.e., the exchange of transversely polarized, real photons). Without the NU-interaction, one still has the basic time-symmetric U-interaction; that is, one has virtual photon exchange but not real photon exchange. As noted above, and as discussed in Kastner (2018) and Kastner and Cramer (2018), the basic probability of the occurrence of the NU-interaction and real photon transfer via the D_1 component is the fine structure constant.

In contrast, traditional quantum field theory (QFT) uses the entire D_F universally. In view of the distinct physical significance of the real and imaginary part of the Feynman propagator as noted above, which holds regardless of the specific model considered, a shortcoming of traditional QFT is that no physical distinction can be made in that theory between radiative and nonradiative processes at the level of the propagator. Indeed, in standard QFT the term "virtual photon" is routinely taken as synonymous with an internal line in a Feynman diagram. This is inadequate and misleading, as it is only a contextual criterion (depending on "how far out we look") and thus does not describe the photon itself. While Davies' definition quoted above – virtual photon is off the mass shell and corresponds to the time-symmetric propagator, while real photon is on the mass shell and corresponds to the free field term – is the correct account of the physical distinction between real and virtual photons, his treatment of the real/virtual distinction in both Davies (1971) and (1972) falters into an ambiguous one alternating between (1) the standard, inadequate QFT characterization of the real versus virtual distinction as a merely contextual one, that is, as an "internal" versus "external" photon dependent on our zoom level, and (2) the mistaken assumption that a real photon must have an infinite lifetime and therefore can only be truly external.[11] In particular, he appears to apply the uncertainty principle to the lifetime of real photons. However, a real photon obeys energy conservation, and its

---

[11] Davies notes that real photons are massless, but this just means they do not decay into other quanta. It does not preclude them from being emitted and absorbed. However, Davies does correctly criticize Feynman's purely contextual account of the "real vs. virtual" distinction by noting that a true virtual photon has no well-defined direction of energy transfer and is described by the time-symmetric component of D_F (i.e., the real part $\bar{D}$) only (Davies, 1972, p. 1028).

lifetime is therefore not limited in that way.[12] The fact that real photons are *both*
emitted and absorbed and therefore can be considered a form of "internal line" is
key in understanding the relevant quantum analog of the LTB condition. Indeed,
all real photons are "internal" in the QDAT, since real photons can only be created
through the participation of both emitters *and* absorbers.

So, keeping in mind that it is indeed possible to have a "real but internal"
photon, let us review another useful account given in Davies (1971) of the relevant
LTB condition for the QDAT. Davies correctly notes that the fully quantum form
of the LTB is simply the requirement that there are no transitions between *external*
fermion/photon states $|\beta\rangle = |\psi, n\rangle$ where photon number $n \neq 0$. He writes this as:

$$\sum_{\beta'} |\langle \beta' |S|\alpha\rangle|^2 = 0 \qquad (5.15)$$

where $|\alpha\rangle$ are states with $n = 0$ and $|\beta'\rangle$ are states with $n \neq 0$. This is in keeping
with the theorem (5.12) and the discussion of (5.13) above. But of course, the
transition probability for each value of β' is a non-negative quantity, so each term
must vanish separately:

$$|\langle \beta' |S|\alpha\rangle|^2 = 0, \forall \beta'. \qquad (5.16)$$

Also, note that by symmetry the restriction on external photon states $n \neq 0$ holds
for both initial states and final states. That is, one must exclude transitions *from*
states $|\alpha'\rangle$ as well as transition *to* states $|\beta'\rangle$. *Thus, the QDAT describes a world in
which there simply are no truly external photons.* This, of course, simply corres-
ponds to setting the independent quantized electromagnetic field A_μ to zero.

Again, this does *not* mean that real photons are disallowed, an inference that
leads to confusion in Davies' account. As emphasized above, in the QDAT, the
only way one obtains a real photon at all is through *both* emission and absorption,
that is, the participation of both the emitter and absorber(s) in the NU-interaction.
The creation of the real photon field can be quantified in terms of a complete set of
field components propagating between the emitter and absorbers; this has been
presented in Kastner and Cramer (2018) and is reviewed below. In effect, the
generation of a complete set of emitter/absorber fields with an appropriate phase
relationship is the entire content of the quantum LTB condition.

---

[12] Even if one disputed this, it is well known that emission and absorption require a finite spread in the energy
level. So one cannot argue that a photon has an infinite lifetime because it has a definite energy; it is possible for
a real photon to have a spread in energy. Davies himself says of the Feynman propagator D_F in (5.14): "The real
part $[\bar{D}]$ gives rise to the self-energy and level shift, whilst the imaginary part $[D_1]$ gives the level width, or
transition rate for real photon emission" (Davies, 1972, p. 1027). Here, Davies uses the customary term "self-
energy," but in fact, no real energy is conveyed by the time-symmetric propagator; it only conveys force, so the
term "self-action" is more accurate. This point is elaborated in note 16 and in Chapter 9.

Davies views the existence of the D_1 term in the context of the restriction (5.15) as paradoxical, since he identifies the term "real photon" solely with an external photon.[13] If we let go of that restriction (as was justified above in our observation that a real photon can indeed be emitted and absorbed), we find that real photons are indeed transferred between currents via the D_1 term. In fact, Davies (1972) gives a quantitative account of how this occurs (although he hesitates to acknowledge those "internal" photons as real photons, calling the relevant construction "formal"). We now review that account.

First, Davies notes the property

$$D^+(x-y) = -D^-(y-x), \tag{5.17}$$

which is useful in what follows. Looking again at the expression from (5.12) for the first-order interaction,

$$\frac{1}{2}\sum_{i,j}\int j_i^\mu(x)D_F(x-y)j_{\mu,j}(y)\,d^4x\,d^4y. \tag{5.18}$$

This is the first-order term in the S matrix, corresponding to the exchange of one photon (either virtual or real, since D_F does not make this distinction). Using the decomposition (5.14) for D_F, we have:

$$\frac{1}{2}\sum_{i,j}\int j_i^\mu(x)\left(\bar{D}(x-y) - \frac{i}{2}D_1(x-y)\right)j_{\mu,j}(y)\,d^4x\,d^4y. \tag{5.19}$$

As Davies notes, the first term (real part) gives us the basic time-symmetric interaction corresponding to off-shell (virtual) photons, while the second term (imaginary part) corresponds to on-shell, real photons. The imaginary part can be written in terms of (5.8) as:

$$\frac{1}{4}\sum_{i,j}\int j_i^\mu(x)(D^+(x-y) - D^-(x-y))j_{\mu,j}(y)\,d^4x\,d^4y, \tag{5.20}$$

which using property (5.17) becomes

$$\frac{1}{4}\sum_{i,j}\int j_i^\mu(x)(D^+(x-y) + D^+(y-x))j_{\mu,j}(y)\,d^4x\,d^4y. \tag{5.21}$$

[13] Davies (1972, p. 1027) suggests that real photons can interfere with virtual photons, resulting in cancellation of the advanced effects of a real photon (which he assumes has an infinite lifetime). But this is only a semi-classical argument that does not carry over into the fully quantum form of the DAT, since different photons do not mutually interfere, and certainly not photons with different physical status regarding whether or not they are on the mass shell. This is also evident from the form of (5.16), in which different external photon states must vanish separately. Davies appeals to a semi-classical argument because he doesn't acknowledge that one can have a real, but "internal," photon.

Because of the double summation over i, j, the two terms are the same, so we are left with:

$$\frac{1}{2} \sum_{i,j} \int j_i^\mu(x) D^+(x-y) j_{\mu,j}(y) \, d^4x \, d^4y. \tag{5.22}$$

In other words, the Feynman propagator leads to the radiation of positive frequencies only. (The opposite phase relationship between the fields generated by emitters and absorbers would lead to the Dyson propagator, with negative frequencies being radiated.)

Now, the final step is to note that D^+ in the integrand of (5.22) factorizes into a sum over momenta, that is,

$$D^+(x-y) = i\langle 0|\hat{A}(x)\hat{A}(y)|0\rangle = i \sum_k \langle 0|\hat{A}(x)|k\rangle\langle k|\hat{A}(y)|0\rangle. \tag{5.23}$$

Again, this represents a real photon, since the action of each of the creation and annihilation operators in A is to create and to annihilate a real, on-shell photon in mode k. But the photon can only end up going to one absorber, not to the many different absorbers implied by the sum, so this is why there has to be "collapse" or reduction, with an attendant probability for each possible outcome. Again, we only get this factorizable "free field" in the presence of the NU-interaction (or what is called "absorber response" at the non-relativistic level). Thus, quantization arises not from a preexisting system of oscillators, but rather from a specific kind of field interaction, that is, the NU-interaction. Note that the right-hand side of (5.23) describes a sum over products of conjugate transition amplitudes for states of varying momenta.[14] This reflects the fact that a real photon is not really a Fock state, which designates only an offer wave, but is really a squared quantity – essentially the vacuum expectation value of a projection operator. Thus, we clearly see the physical origin of the squaring procedure of the Born Rule: the photon is created through an interaction among emitter *and* absorbers (not unilaterally), but ultimately can only be delivered to one of the absorbers. In addition, this product form gives us the correct units for the photon, that is, energy, whereas the units corresponding to a ket alone are the square root of energy.

In light of (5.23), the quantum version of the "light-tight box" condition is simply the completeness condition; that is, the factorization over quantum states of a transferred photon can only be carried out if the set of states is complete.

---

[14] The two amplitudes have different spacetime arguments, so there is an overall phase factor reflecting the emission and absorption loci with respect to the inertial frame in which the fields are defined. The photon itself has no inertial frame and is described only by the conserved currents it transfers, corresponding to the square of the field amplitude. Thus the phase factor applies to the fermionic field sources. The different roles of photons and fermions in RTI are discussed further in Chapters 6 and 8.

Physically, this means that absorbers corresponding to each possible value of k must respond or, more accurately at the relativistic level, that the emitter and absorbers must engage in a mutual interaction, above and beyond the off-shell time-symmetric field $\bar{D}$, to generate an on-shell field that can be factorized, corresponding to the quantum completeness condition.

There is a bit of a subtlety here in understanding what counts as a "complete set" of momenta. Typically, one assumes a continuum of momentum values, but this is a mathematical idealization that does not apply to physically realistic situations, and in particular not to the QDAT. All that is required is that all momentum projectors $|k_i\rangle\langle k_i|$ for the fields exchanged between the emitter and absorbers $i = \{1, N\}$ sum to the identity. A particular k_j refers to a particular absorber j that engages with the emitter to jointly create one component of the on-shell field whose quantum state can be written as $|\Psi\rangle = \sum_i \langle k_i|\Psi\rangle|k_i\rangle$. Thus, these states $|k_i\rangle$ have finite spread corresponding to the effective cross-section of each absorber and any uncertainty in the photon energy.

Even though all N absorbers contribute to create the on-shell field, as noted above, the real photon can ultimately be received by only one absorber, and this corresponds to non-unitary state reduction to the value k_j for the received photon, with the probability $|\langle k_j|\Psi\rangle|^2$ (the Born Rule). Thus, besides the elimination of the independent system of field oscillators represented by the quantized field $\hat{A}$, the entire content of the quantum LTB is just the quantum completeness condition and the phase relationship that selects the Feynman rather than Dyson propagator.[15]

5.2.6 Non-unitarity

The S-matrix is unitary if all interacting currents are included in the sum (5.12) such that all state transitions involving those currents start from the photon vacuum state and return to the photon vacuum state. In this case, the net "free field" vanishes because of the QDAT condition disallowing truly unsourced photon states (5.15). However, for a subset of interacting currents, the S-matrix contains a non-unitary component: that of the "free field" D_1. While Davies (1972) found this feature "puzzling," the present author has noted that this element of non-unitarity

---

[15] The two choices of phasing of absorber response reflect the fact that the theory has two semi-groups. These are actually empirically indistinguishable. For the Feynman propagator, bound states are built on positive energies; for the Dyson propagator, bound states are built on negative energies. Thus, any observer would see an arrow of time/energy pointing to what they would consider the "future," and what constitutes "positive" or "negative" energy is only a convention based on the structure of the bound states. Here we differ with Davies (1972, pp. 1022–24), who suggests that the two choices are not the time-inverse of one another. That conclusion follows only if one retains the positive-energy structure of bound states while employing the Dyson propagator. But, arguably, that is not appropriate.

provides a natural account of the measurement transition (Kastner, 2015; Kastner and Cramer, 2018).

The non-unitary property of the S-matrix in a vacuum-to-vacuum transition for a subset of all interacting currents is also discussed by Breuer and Petruccioni (2000, pp. 40–41). In a study of decoherence, these authors take note of the fact that the Feynman propagator is complex and contains an imaginary component of the action based on the effective "free field" D_1. For a single current, the vacuum-to-vacuum scattering amplitude $S(D_1)$ corresponding to this component is

$$S(D_1) = \exp\left(-\frac{1}{4}\int j^\mu(x)D_1(x-y)j_\mu(y)d^4xd^4y\right). \tag{5.24}$$

The integral in the exponential is real and positive, and can be interpreted as half the average number of photons $\bar{n}$ emitted by the current (and absorbed by another current). The vacuum-to-vacuum probability associated with the free photon field is

$$|S(D_1)|^2 = e^{-\bar{n}} < 1, \tag{5.25}$$

which corresponds to the probability that no photon is emitted by the current (note that if $\bar{n} = 0$, the probability is unity). Note that this is an explicit violation of unitarity at the level of the S-matrix for a single current (i.e., when final absorption of the emitted photon(s) by other current(s) is not taken into account). Based on this result, Breuer and Petruccioni note that it is the D_1 component that leads to decoherence. The present author discusses the crucial dependence of decoherence on non-unitarity in Kastner (2020a).

Davies further notes that the complement of (5.25) is the probability of photon emission by the current:

$$1 - |S(D_1)|^2 = 1 - e^{-\bar{n}} = \sum_{m=1}^{\infty} e^{-\bar{n}}\frac{\bar{n}^m}{m!}, \tag{5.26}$$

where each term in the sum is the probability of emission of m photon(s), the Poisson distribution applicable to the well-known infrared divergence.

To recap what has been covered in Section 5.2: the so-called light-tight box (LTB) condition applying to the direct-action theory of fields (DAT) needs critical review and requires explicit revision at the quantum level. The condition at the classical level, (5.3), can be interpreted to mean that there is no truly unsourced radiation field, rather than the usual interpretation that "all emitted radiation is absorbed," since the condition actually says nothing about absorption, but says only that the net free field is zero. At the quantum level (QDAT), the condition is represented by (5.16), which simply says that there exist no true "external" photon states. A theorem showing the equivalence between the standard quantized field theory and the QDAT reveals that the condition is simply the vanishing of the quantized field A_μ. Instead, in the QDAT, interactions are mediated by a non-quantized electromagnetic potential that directly connects charged currents through the time-symmetric propagator.

In order to understand the conditions for real photon generation in the QDAT, it must be understood that a real, on-shell photon can indeed be emitted and absorbed and therefore be "internal," with a finite lifetime. Under a form of the quantum completeness condition, and governed by the fine-structure constant (and relevant transition probabilities), an effective "free field" corresponding to the even homogeneous solution, D_1, can arise. This is the quantum analog of "absorber response," which at the relativistic level is a mutual non-unitary interaction between emitter and absorber(s) that gives rise to one or more real, on-shell photons, even though such photons are technically "internal." The presence of D_1 converts the time-symmetric propagator into the usual Feynman propagator (Equation (5.10)). No "light tight box," that is, no particular configuration of absorbers, is required for these processes to occur, so that no particular cosmological conditions need obtain in order for the QDAT to be fully applicable.

5.3 The Micro/Macro Distinction

The identification of the fine structure constant (square of the charge or coupling amplitude) as the basic probability for the NU-interaction allows an unambiguous fixing of the allegedly "shifty split" between the quantum realm of microscopic entities (such as atoms) and the classical realm of macroscopic objects. This is because the NU-interaction is the measurement transition, and the occurrence of a measurement transition is what defines the "split."

First, some additional terminology will be useful. Let us call quantum sources of photons, such as atoms, "micro-emitters" and "micro-absorbers" to distinguish them from macroscopic sources and sinks. A macroscopic sink, or macro-absorber, is a generic laboratory detector, that is, a device that can reliably detect an emitted quantum in the appropriate state. In other words, a photon in a particular prepared state will be detected by this device with virtual certainty in a given unit time appropriate for scientists in a laboratory. Such a detector, or macro-absorber, is a bound system of many micro-absorbers. Now suppose the number of its constituent micro-absorbers is N. In order for it to detect an emitted photon in the appropriate state reliably, at least one out of its N constituent micro-absorbers must respond (more precisely, participate with the emitter in an NU-interaction) within the desired unit time with certainty. In other words, the probability that at least one micro-absorber in this object will respond must approach unity within unit time. How big does N need to be for this?

It turns out that this criterion is met when N approaches the size of objects that we perceive as macroscopic. Consider for example a simple detector D comprising N atoms, and for simplicity assume that they are all in their ground states. These atoms are all micro-absorbers. Now suppose there is a single micro-emitter, an atom in an excited state – call it E. For this simplest of situations, let's set up the experiment so that E can only emit to this single detector D. That is, E is prepared in a state such that any photon it emits will be found at D with certainty. Now, recall that the basic amplitude of CW generation applying to each of the micro-absorbers in D is the elementary charge e (in natural units). Of course, at the relativistic level, this is really the mutual NU-interaction between any one of the atoms in D and the micro-emitter. So the elementary probability of any individual micro-absorber engaging with the micro-emitter in the NU-interaction, which involves both OW and CW generation, is e^2, that is, the fine structure constant $\alpha \sim 1/137$. As noted above, what we need for D to count as a macro-absorber is for at least one of the N atoms in D to generate a confirmation wave with certainty within the relevant unit time.

The relevant probability is most easily obtained by first finding the probability of the complement, namely, the probability that for all N micro-absorbers constituting D, there will be *no* NU-interaction. Let us call this Prob(no NU). For $N = 1$, we find

$$\text{Prob}_{(N=1)}(\text{no NU}) = 1 - a = 0.993. \tag{5.27}$$

So a single atom will not count as a decent detector – it will be very unlikely to trigger the measurement transition. For $N > 1$, the probability that not one of the single micro-absorbers constituting D participates in a NU-interaction is

$$\text{Prob}_{(N)}(\text{no NU}) \sim (1 - a)^N = 0.993^N. \tag{5.28}$$

This quantity decreases with increasing N. If we consider a small but macroscopic detector containing about $N = 10^{23}$ excitable atoms, we get

$$\text{Prob}_{(N=10^{23})}(\text{no NU}) \sim 0.993^{(10^{23})} \sim 0. \tag{5.29}$$

For this small but macroscopic detector, *the probability that not one of the atoms will engage in a NU-interaction with the micro-emitter E is negligible.* This means that at least one micro-absorber in D will engage in a NU-interaction, in which case D has engaged in a NU-interaction (since it does not matter which of D's components is involved).

The virtual certainty that D will engage in a NU-interaction within the appropriate unit time means that it qualifies as a "macro-absorber," since it reliably prompts the measurement transition. This account clearly delineates the micro/macro transition point, meaning that it fixes the so-called shifty split. In fact, there is a split, but it is non-arbitrary, since it is defined by the involvement of a physical system that reliably prompts the non-unitary measurement transition. We see that what allows a system to do this is that it contains a sufficient number of micro-absorbers. This allows us to clearly distinguish the macroscopic world of classical phenomena from the micro-world of quantum objects.

The same analysis permits a definition of the *mesoscopic* level, which involves fairly large and complex systems as compared to elementary particles, but which still retains some quantum features. An example of such a system is a very large molecule called a "Buckyball," which comprises 60 carbon atoms. Mesoscopic objects comprise intermediate-sized numbers of micro-absorbers, so that they have a significant but still far-from-certain probability of engaging in a NU-interaction. For purposes of illustration, suppose that a Buckyball's 60 carbon atoms qualify as micro-absorbers for a given emitter. This gives us a value for Prob(no NU) of

$$\text{Prob}_{N=60}(\text{no NU}) \sim 0.993^{60} \sim 0.66. \tag{5.30}$$

It is quite possible that a Buckyball will engage in a NU-interaction, since Prob (NU) = 1 − Prob(no NU) = 0.34. But it is far from certain. Thus, we see that the probability of participation in a NU-interaction by any particular object, based on the number N of micro-absorbers comprising it, provides a clear quantitative criterion for whether that system qualifies as "macroscopic" (i.e., there is virtual certainty that it will precipitate the non-unitary measurement transition), "meso-scopic" (somewhat likely to precipitate the transition), or "microscopic" (extremely unlikely to precipitate the transition).

The result presented here – that the fine structure constant is the basic probability of generation of the non-unitary measurement transition by an emitter and each of its potential absorbers – is probably the most important of the developments of RTI. It shows that "emitters" and "absorbers" are not primitive notions but are well-defined physical systems whose behavior in connection with measurement can be quantified. It addresses the notorious problem of the "Heisenberg cut" or "shifty split" between the unitary evolution and the non-unitary von Neumann "Process 1" involving the Born Rule. But again, the demarcation is not really a "cut" but is rather a range of values of the number N of constituents of a system, such that the likelihood of the measurement transition occurring at that object becomes greater and greater until it is virtually certain. At that level, the object can be considered "macroscopic."

5.4 Classical Limit of the Quantum Electromagnetic Field

In this section, I discuss how the classical electromagnetic field emerges from the quantum domain through transactions. We first note that so-called coherent states

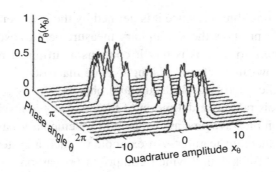

Figure 5.1 Data from photon detections reflecting oscillation of the field source.
Source: https://commons.wikimedia.org/wiki/File:Coherent_state_wavepacket.jpg.

$|\alpha\rangle$ provide the closest correspondence between quantum fields and their classical counterparts. Such states describe an indeterminate number of quanta:

$$|\alpha\rangle = e^{\frac{-|\alpha|^2}{2}} \sum_{n=0,\,\infty} \frac{\alpha^n}{\sqrt{n!}} |n\rangle. \tag{5.31}$$

These states are eigenstates of the field annihilation operator $\hat{a}$; the field in that state does not "know" that it has lost a photon. That is,

$$\hat{a}|\alpha\rangle = \alpha|\alpha\rangle. \tag{5.32}$$

So, in effect, the state describes an effectively infinite and constantly replenished supply of photons. The coherent state can be thought of as a "transaction reservoir" analogous to the temperature reservoirs of macroscopic thermodynamics. In the latter theory, the interaction of a system of interest with its environment is modeled as the coupling of the system to a "heat reservoir" of temperature T. In this model, exchanges of heat between the reservoir and the system affect the system but have no measurable effect on the reservoir. In the same way, a coherent state is not affected by the detection of finite numbers of photons.

Experiments have been conducted in which a generalized electromagnetic field operator is measured for such a state.[16] Detections of photons from a coherent field state generate a current, and that current is plotted as a function of the phase of the monochromatic source, that is, a source oscillating at a particular frequency – for example, a laser (see Figure 5.1). Such a plot reflects the oscillation of the source in that the photons are detected in states of the measured observable (essentially the electric field amplitude) which oscillate as a function of phase (individual photons do not oscillate, however).

---

[16] See, for example, Breitenbach et al. (1997).

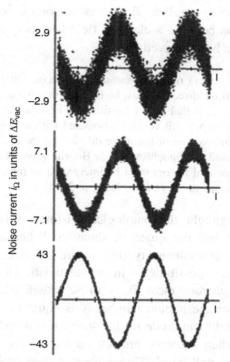

Figure 5.2 Coherent states with increasing average photon number (top to bottom).
Source: http://en.wikipedia.org/wiki/File:Coherent_noise_compare3.png.

The theoretical difference between the quantum versions of fields (such as the coherent state) and their classical counterparts can be understood in terms of the ontological difference between quantum possibilities (e.g., offer and confirmation waves and incipient transactions) and structured sets of actualized transactions. The quantized fields represent the creation or destruction of possibilities, and the classical fields arise from states of the field that sustain very frequent actualized transactions, in which energy is transferred essentially continuously from one object to another. Again, this can be illustrated by the results of experiments with coherent states that "map" the changing electric field in terms of photon detections, each of which is a transaction. For states with small average photon numbers, the field amplitude is small and quantum "noise" is evident (for the coherent state, these are the same random fluctuations found in the vacuum state). As the coherent state comprises larger and larger numbers of photons, the "signal-to-noise ratio" is enhanced and approaches a classical field (see Figure 5.2). Thus, the classical field is the quantum coherent state in the limit of very frequent detections/transactions.

It is the classical, continuous detection/transaction limit, in which the field can be thought of as a classical propagating wave, to which the original

Wheeler–Feynman theory applies. But it is important to keep in mind the fundamental distinction between a classical field and its quantum counterpart. In this regard, Paul Dirac has observed:

Firstly, the light wave is always real, whereas the de Broglie wave associated with a light quantum moving in a definite direction must be taken to involve an imaginary exponential. A more important difference is that their intensities are to be interpreted in different ways. The number of light quanta per unit volume associated with a monochromatic light wave equals the energy per unit volume of the wave divided by the energy hv of a single light quantum. On the other hand, a monochromatic de Broglie wave of amplitude a (multiplied into the imaginary exponential factor) must be interpreted as representing a^2 light-quanta per unit volume for all frequencies. *(Dirac, 1927, p. 247)*

Dirac's comments highlight the ontological distinction between the classical electromagnetic wave and the quantum situation. Whereas the classical wave conveys energy through its intensity (the square of its electric field strength), quantum states represent possibility – in the relativistic case of a coherent field state, the *number of photons* most likely to be actualized. The amplitude of a coherent state with average photon number N is equal to $\sqrt{N}$ (which is proportional to the electric field amplitude for the state); it is a multi-quantum probability amplitude that, when squared, predicts that the most probable number of photons to be detected will be N. Thus, coherent state probabilities address the question: "How many photons will be actualized?" – and it is to this question that the squared amplitude $(|\alpha|^2)$ of the coherent state $|\alpha\rangle$ applies. In contrast, the squared amplitude of the classical wave addresses the question, "What is the energy associated with the actualized photons?" The *energy* $E = hv$ of a particular actualized (detected) photon is frequency-dependent, but the *probable number* of actualized photons is not.

Yet the unity of the two descriptions is still expressed in the fact that it is not the classical field that really conveys energy; rather, it is the *intensity* (squared amplitude) of the field. This can again be traced to the underlying transactional description. A photon does not exist in spacetime unless there is an actualized transaction involving an offer wave and a confirmation wave, which is what is described by the squaring process (Born Rule). Energy can only be conveyed by a detected photon, not by an amplitude (offer wave) only. This fact appears at the classical level and can be seen as a kind of "correspondence principle" between the two descriptions.

Some further remarks are in order regarding the ontology of coherent states in the transactional picture. A coherent state, as a support of the Coulomb field, is a virtual photon entity, that is, not really an "offer wave," which strictly speaking corresponds only to a radiated photon. Recall that in the direct-action theory, the Coulomb field is not quantized and is a measure of virtual photon activity only.

At the quantum level, Faraday's "lines of force" are not actualized entities that exist in spacetime, but are constructs that specify the forces on charged objects such as electrons. Such forces fundamentally act at the virtual photon level, that is, at the level of unitary evolution. It is only when the force is actualized, that is, acts through a distance to exchange energy (in a transaction) that we may detect the phenomena that allow us to measure the force.

So when one defines a coherent state $|a\rangle$, one is really describing a system of excited photon sources (such as a laser) and absorbers that together yield a well-defined probability for transfer (radiation) of a number n of real photons at any particular time t. Thus, in RTI, a "coherent state" is *not* the state of a preexisting quantized electromagnetic field as is customarily assumed. Rather, it describes the sources of the field. Any OW and CW that occur are always for well-defined photon numbers; that is, they are Fock states. The time for generation of any particular Fock state $|n\rangle$ is always fundamentally uncertain (corresponding to phase uncertainty of the Fock state). But a coherent source, such as a laser, collectively gives rise to a temporal phase relation, since as a whole, it is not committed to any particular Fock state $|n\rangle$ at any time t. It is only committed to transfer *some* Fock state $|n\rangle$, with a Poisson probability $|\langle n|a\rangle|^2$, at time t.

5.5 Nonlocality in Quantum Mechanics: RTI versus rGRWf

GRW approaches were briefly reviewed in Chapter 1. The most recent version of GRW is a proposal by Tumulka, "relativistic GRW flash" or rGRWf, which attempts to provide a relativistically compliant version of that approach, together with a so-called flash ontology that provides for specific measurement results without depending on a problematic compression of the wave function.[17] This section argues that RTI does a better job of accommodating relativity.

5.5.1 Gisin's Result

Gisin (2010) has recently argued that under certain conditions, and assuming strong causality (i.e., an event can only influence other events in its future light cone), Bell's Theorem will rule out the ability of all hidden variables (whether local or nonlocal) describable by a covariant probability distribution to reproduce the nonlocal correlations between spacelike detectors for EPR-type entangled states. Specifically, Gisin considers the usual "Alice and Bob" EPR situation, and defines Alice's and Bob's results, α, β, respectively, as functions F_{AB} [F_{BA}] of their

---

[17] One such problem is that a sudden compression of the wave function in the position basis results in an essentially infinite range of energies for the particle.

measurement settings $\vec{a}$, $\vec{b}$ and the value of some nonlocal hidden variable λ. The order of the subscripts on F indicates which measurement is first in the frame considered. Thus if Alice measures first, her outcome $\alpha = F_{AB}(\vec{a},\lambda)$; if Bob measures first, his outcome $\beta = F_{BA}(\vec{b},\lambda)$. Gisin then constructs the analogous function S for the outcome measured second, and notes (assuming time-asymmetric strong causality) that it must also be a function of the measurement setting for the first measurement, that is, $\beta = S_{AB}(\vec{b},\vec{a},\lambda)$. Analogous expressions are constructed in the frame in which Bob measures first. Gisin then notes that, if covariance holds, the same λ should characterize the results irrespective of the frame considered, so that we must have

$$\alpha = F_{AB}(\vec{a},\lambda) = S_{BA}(\vec{b},\vec{a},\lambda) \tag{5.33a}$$

and

$$\beta = F_{BA}(\vec{b},\lambda) = S_{AB}(\vec{a},\vec{b},\lambda) \tag{5.33b}$$

but there is no λ that can satisfy (5.33a, b), since they actually imply that λ is a local variable and these are already ruled out by Bell's Theorem. Thus, Gisin has ruled out the ability of nonlocal hidden variables to yield a covariant account of actualized outcomes for quantum-correlated spacelike events. This formalizes observations such as Maudlin's (1995) that Bohmian-type "preferred observable" accounts seem to be at odds with relativity.

However, as noted, Gisin's analysis presupposes "strong causality." That is, it specifies which observer's outcome was prior to the other observer's outcome, with the assumption that the second observer's result depends on the setting and outcome of the first observer. Thus, his result does not rule out the ability of time-symmetric approaches to yield a covariant account. Indeed, we will see that RTI can provide all the benefits of Tumulka's GRW "flash ontology" model, "rGRWf" (2006), without being a modification of quantum theory.

5.5.2 Is There Really a GRW Advantage?

I should first address the claim sometimes made that GRW has an advantage over TI in that the former spells out a particular measurement result, while TI's offer/confirmation wave encounter does not (strictly speaking, the latter determines a *basis* for the determinate outcome while not specifying which one occurs[18]). But arguably, this advantage of GRW is only illusory. The GRW outcome is specified by resorting to an ad hoc and physically undefined (in terms of any existing theory)

---

[18] It thus gives a physical explanation for the projection postulate of standard QM, as shown in Chapter 3.

"flash" process. The worst that one can say at present concerning TI (and RTI) is that there is no mechanistic causal story behind the realization of a particular transaction (outcome) as opposed to a competing "incipient" one, which makes it at least no worse off than GRW in terms of providing concrete physical reasons for a specific measurement result. Meanwhile, TI does give a clear account of the measurement process in terms of absorption, as discussed above.

A common point of confusion concerning TI is the failure to recognize that confirmations are generated for *all* components of the offer wave for which absorbers are present, resulting in a weighted set of incipient transactions corresponding to von Neumann's "Process 1" (or the projection postulate). This set of incipient transactions corresponds to an "ignorance"-type mixture, in that measurement has definitely occurred and the uncertainty concerning outcome is epistemic. The realization of a particular transaction out of a set of incipient ones can be seen as a kind of spontaneous symmetry breaking, as discussed in Chapter 4. So it would not be fair to claim, as some have done, that TI does not provide a complete account of measurement.

5.5.3 A Dilemma Reexamined

Tumulka has argued that, in his words, "Either [1] the conventional understanding of relativity is not right, or [2] quantum mechanics is not exact."[19] But this particular dilemma needs to be examined more closely, as horn [1] has more content than is customarily assumed. By [1], Tumulka has in mind the usual assumption that any exact, realist interpretation of quantum theory must involve a preferred inertial frame or "spacetime foliation." But as noted above, there is something more to be questioned in the "conventional understanding" of relativity: an inappropriately strong time-asymmetric causality constraint. So horn [1] really has two different options: [1a] "there is a preferred frame" or [1b] "causal influences can be time-symmetric." Thus option [1] can be chosen *without* embracing a preferred frame, in the form of [1b]. That is, one can reject the necessity of a preferred frame and argue that what is "not right" about the conventional understanding of relativity is the notion that it rules out time-symmetric influences.

Whereas GRW "spontaneous localization" approaches such as Tumulka's "rGRWf," in an effort to avoid the preferred foliation that is assumed to be the only option contained in [1], choose [2] and modify quantum theory in an explicitly ad hoc manner, RTI chooses [1], but *not* in the sense of [1a] involving a preferred foliation as is usually assumed. Instead, it is noted that relativistic restrictions

[19] Tumulka (2006, p. 352).

should be properly considered to apply only to in-principle observable events, and that subempirical causal time symmetry – in the sense of our not being constrained to a choice of which of two events is the "cause" and which the "effect" – should be accepted via option [1b].

Indeed, a similar relaxation of strong causation is just what Tumulka adopts in order to argue that the nonlocal correlations arising between spacelike separated flash events in his model do not violate covariance. He remarks: "An interesting feature of this model's way of reconciling nonlocality with relativity is that the superluminal influences do not have a direction; in other words, it is not defined which of two events influenced the other."[20] Note that, since these are spacelike separated events, there is a frame in which one is first and a different frame in which the other is first, so one could argue that there can be time-reversed causal effects in one frame or the other, depending on which event is arbitrarily considered the "cause" and which the "effect." (One might object here that Tumulka addresses this by saying that no such causal order exists, but that is precisely the case in RTI as well.) So we see the relativistic version of GRW already heading in the direction of time symmetry, or at least toward weakening the overly strong "causality" assumption so often presumed in the literature.

Under RTI, sets of possible transactions (whose weights, interpreted as probabilistic propensities, are reflected in the Born Rule) provide a covariant distribution of possible spacetime events. Moreover, there is nothing about the sets of actualized events in RTI that can be seen as noncovariant, as in the actualized events discussed by Gisin. This is because, under RTI, it is not assumed that the events (Alice's and Bob's outcomes) had a strict temporal causal order. Gisin's observation regarding the noncovariance of actualized events does not apply to sets of actualized events in RTI, since all events are dependent on both the emitter's "offer wave" and the absorber(s)' "confirmation wave(s)." Just as in Tumulka's account of his nonlocally correlated flashes, there is no need (nor would it be appropriate) to define which of a set of spacelike separated events is the "cause" and which is the "effect" of a particular outcome. The emitter and absorber(s) participate equally and symmetrically in the transaction leading to the outcome(s). Thus, actualized transactions play the part of the "flashes" in Tumulka's model, but without the necessity of modifying the dynamics of quantum theory. While Tumulka has opted for a modification of quantum theory in order to avoid a preferred frame – our [1a] above – he has also made use of [1b] which, in view of the time-symmetric alternative of RTI, obviates the need for modifying quantum theory in the first place.

In the next chapter, we consider some challenges to TI that are fully resolved, as well as some specific applications of the transactional picture.

---

[20] Tumulka (2006, p. 352).

6

Challenges, Replies, and Applications

6.1 Challenges to TI

Tim Maudlin considered TI in his book *Quantum Nonlocality and Relativity* (2002, pp. 199–201), which explored the apparent tension between quantum theory and relativity in terms of nonlocal effects and influences. He concluded at that time that TI was not viable based on a type of thought experiment which seemed to imply an inconsistency. Maudlin's challenge and similar challenges have been addressed by several authors, who have argued that it is not fatal for TI.[1] The present author is among those who have argued that Maudlin-type challenges are not fatal. However, the basic concern behind them was a worthy one that prompted further development of the interpretation. The latest development at the relativistic level reveals that in fact the Maudlin challenge cannot be mounted at all for massive quanta. A version corresponding to photons remains, but as we will see, it is not a problem.

First, some preliminary remarks. As discussed in Chapter 4, a key component of this development of "possibilist TI" (PTI)[2] is that offer and confirmation waves are *physical possibilities* that are subempirical and pre-spatiotemporal. Another component is the necessity to embrace a "becoming" view of events rather than a "block world" view. (The latter will be more fully examined in Chapter 8.) Offers and confirmations should be thought of not as propagating within spacetime (in either temporal direction), but rather as acting instead at a pre-spacetime level, in what we termed the *quantum substratum* in the previous chapter. Actual spacetime events are emergent from the transactional process; they are supervenient on that process rather than being present a priori as part of a spacetime substance or "block world," as is assumed in Maudlin-type challenges. While the PTI ontology –

---

[1] Berkovitz (2002); Cramer (2005); Marchildon (2006); Kastner (2014a).
[2] In this book I use "PTI" and "RTI" interchangeably based on context (whether ontological or formal); they name the same model.

especially the subempirical, extra-spatiotemporal nature of the offers and confirmations – has been viewed with some initial skepticism, it should be kept in mind that most competing interpretations incorporate subempirical features as well. For example, the Everettian or "many worlds interpretation" assumes a subempirical, extra-spatiotemporal splitting of worlds or observers, and the DeBroglie/Bohm theory assumes a subempirical, extra-spatiotemporal "guiding wave" which is conceptually very similar to TI's offer wave. Because the Hilbert space structure of the theory is not reducible to that of spacetime – the manifold of empirical events – any realist interpretation of quantum theory must acknowledge that the mathematical formalism refers (at least in part) to something transcending the empirical realm.[3] This inevitable message of the theory is again reflected in Bohr's comment that quantum processes "transcend the spacetime construct."

6.1.1 The Maudlin Challenge: Introduction

The Maudlin challenge is a critique of the "pseudotime" account presented in Cramer (1986), in which transactions are established in a forward-and-backward temporal process between an emitter and a set of absorbers. It proposes a thought experiment in which the placement of a distant absorber for a possible transaction is contingent on the failure of a competing transaction with a nearby absorber. Basically, it is a critique of the idea of absorbers as a static backdrop for the "competition" among incipient transactions.

The basic argument can be summarized as in Figure 6.1. A source emits massive (and, therefore, Maudlin assumes, slow-moving)[4] particles to either the left or the right, in the state $|\Psi\rangle = \frac{1}{\sqrt{2}}[|R\rangle + |L\rangle]$, a superposition of "rightward"-and "leftward"-propagating states. OW components corresponding to right and left are emitted in both directions, but in this arrangement, it's assumed that only detector A can initially return a CW (since B is blocked by A). If the particle is not detected at A (meaning that the rightward transaction failed), a light signal is immediately sent to detector B, causing it to swing quickly around to intercept the particle on the left. B is then able to return a CW, but it is only of amplitude $\frac{1}{\sqrt{2}}$ and yet the particle is certain to be detected there, which Maudlin claims is evidence of

---

[3] An interesting image reflecting this mathematical fact can be found on the cover of Bub's *Interpreting the Quantum World* (1997). The cover image shows M. C. Escher's famous print *Waterfall*, depicting a scene with a physically impossible topology (i.e., one that could not actually fit into spacetime). Three separate areas of the print are highlighted, and each of these could exist in isolation in spacetime, but the global connections between them cannot. In the interpretation proposed herein, the smaller highlighted "normal" areas represent the actualized transactions, and the larger shaded area of topologically "impossible" global interconnections belong to the pre-spacetime realm of possibility (i.e., offer and confirmation waves).

[4] When one considers the relativistic level, it turns out that particles with nonvanishing rest mass are not subject to transactions by way of matching confirmations. They participate only indirectly in transactions. It is only photons that are actualized directly through OW and CW. This is discussed in Section 6.1.2.

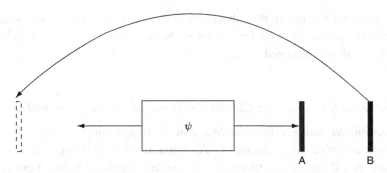

Figure 6.1 The Maudlin contingent absorber experiment.

inconsistency on the part of TI. He also argues that the "pseudotime" picture cannot account for this experiment, since the outcome of the incipient transaction between the emitter and the nearby absorber must be decided supposedly without a CW from the more distant absorber. However, as we have seen in Chapter 5, if there is a transactional process – a NU-interaction – it is only because there is a complete set of absorber responses. That is, a necessary and sufficient condition for the transactional process constituting measurement is a complete set of absorber responses. If any such process occurs in this scenario, it is because some other background absorber in the lab corresponding to the leftward direction responded. So this aspect of the Maudlin objection fails at this point, in any case.

There are quite a few variations on the original Maudlin challenge. Miller (2011) has proposed a version involving photons. This version envisions a photon OW split by a half-silvered mirror into two beams A and B; the beam in B is detoured by a fixed set of mirrors to delay its absorption by detector B. If it is not detected at A at $t = 1$, a movable mirror is quickly inserted into the beam going to detector B, with the idea being that the OW component in that arm is diverted to detector B′ (perhaps with different properties such as a polarization filter).

This scenario fails because it applies classical notions to a highly nonclassical quantum state of light. In fact, it is a feature of standard quantum theory that the photon cannot be localized along its path in this way.[5] Photon localization is essentially equivalent to trying to determine the arrival time of a photon at a detector, which is a highly nontrivial issue. In fact, any actualized photon corresponds to an essentially monochromatic Fock state. The time of arrival of such a photon is maximally uncertain according to the uncertainty principle. This

---

[5] The notorious problem of photon localization is discussed at length by, for example, Bialynicki-Birula (1996) and Mandel (1966). One can construct theoretical approximately localized position "wave packets" for photons, but these do not coincide with the energy density of the photon's field, so there is no consistent way to describe a photon as localized. See Saari (2011, p. 51) for a discussion. Ultimately, this is because a real photon simply is not localized.

means that even if its path in the laboratory is subject to delay as in this scenario, it can effectively arrive in zero time relative to any inertial frame, which puts it always ahead of any material object.

6.1.2 The Maudlin Challenge Cannot Be Mounted for Particles with Rest Mass

In this section, we will see that the Maudlin challenge simply evaporates for any particle with non-vanishing rest mass.[6] Any quantized field theory can be formally reexpressed as a direct action theory, as shown by Narlikar (1968). However, as a contingent matter of the field interactions in our world, only photons (in virtue of being massless gauge bosons) participate directly in the NU-interaction described in the previous chapter. The main reasons for this are (1) offers and confirmations are elementary field excitations, and (2) fermions are matter particles that serve as sources of the vector bosons that mediate forces and correspond to spacetime symmetries. Only an elementary massless force mediator can participate in the transactional process (the NU-interaction). The massive gauge bosons (Z and W+/−) have complex mass-shell energies; they are unstable and decay quickly into other particles. Their inherent complexity makes them ineligible for the NU-interaction, which involves the transfer of real-valued energy. Gluons are massless, but participate in the strong interaction, which keeps them confined to bound states.

A bound system such as an atom or composite particle is not an excitation of any particular individual field. It is the result of an interaction among field excitations. Since it is not a field excitation itself, it is not subject to the direct action model, which applies only at the level of individual field excitations. Elementary fermionic field excitations, such as electrons, act as emitters and absorbers and are not themselves exchanged via transactions. As sources for the photon NU-interaction, they are actualized indirectly, in that they are correlated with the actualized photon states. This is discussed in more detail below and in Section 6.5.

The relevance of the foregoing for the Maudlin challenge is that it depends on the idea of a "slow-moving offer wave." But since the only objects that qualify as offers and confirmations are photons, there is no such entity. We saw above that delaying the photon's path by bouncing it back and forth off mirrors does not make the photon into a real "slow-moving offer wave," since the photon is not localizable. In what follows, we'll examine more closely why an elementary fermion such as an electron, though formally eligible to count as an offer wave, cannot instantiate the Maudlin scenario either.

[6] This section is based on material in Kastner (2019a).

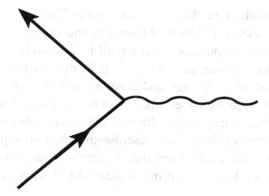

Figure 6.2 QED vertex.

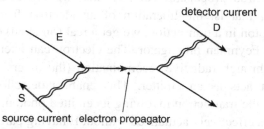

Figure 6.3 Electron detection.

Fermionic matter sources have a very different physical character from the bosonic fields of force to which they give rise. We can see this asymmetry clearly in the basic quantum electrodynamics (QED) vertex (Figure 6.2). It has only one photon line, but two fermion lines: one incoming and one outgoing. This reflects the structure of the coupling between the Dirac field and the electromagnetic field, given by the term $(-eA_\mu\bar{\psi}\gamma^\mu\psi)$ from the QED Lagrangian. In the diagram, A_μ represents the electromagnetic field (photon), while ψ and $\bar{\psi}$ represent the incoming and outgoing electron currents. As noted in the previous chapter, the charge e is the coupling amplitude.

In Figure 6.3, an electron is liberated from a bound state by absorbing a photon from some other charged current S in a transaction (indicated by the double wavy lines). It then engages as an emitter in a transaction with another charged source field, typically an electron, in detector D. (Recall from the previous chapter that at the quantum relativistic level, photon OW and CW are mutually generated in the NU-interaction.) The outgoing electron acquires a well-defined state corresponding to the actualized emitted photon state, though the electron is not involved in its own NU-interaction (i.e., there is no "electron confirmation").

Thus, a fermion such as an electron is not "emitted" or "absorbed" in the same way as a photon. Rather, it is *liberated* from a bound state (by absorbing energy) and can exist as a free, autonomous entity until it is possibly incorporated into a new bound state (by emitting energy). This is in distinct contrast to a photon, which is always tied to its emitter and absorber and is never autonomous. An emitter or absorber such as an electron or atom acquires a determinate state indirectly through photon transactions for which it serves either as an emitter or as an absorber. Born probabilities still describe the electron's possible outgoing states, but they arise indirectly from the electron's participation in the photon transactions, which mediate the electron's detection. This issue is elaborated in Section 6.5.

The asymmetry between a fermionic source of the electromagnetic field, such as the electron E, and its emitted and absorbed fields (the photon lines), is evident in Figure 6.3: an electron propagator connects the two interaction vertices. In an ionization process, that is, the liberation of an electron from an atom by its absorption of a photon in a transaction, we get a real (on-shell) electron, described by the pole in the Feynman propagator. The electron can later become part of a new bound state through radiative recombination (the inverse of ionization). In such a process, it acts as an emitter. This enables detection of the electron indirectly, through the transaction involving its emitted photon. Thus, an outcome for the electron E is effectively actualized without E having participated directly in an "electron transaction"; rather, E has given rise to an emission event, which is now part of the spacetime manifold.

Between its liberation and reincorporation in a new bound state, the free electron can be subject to unitary interactions, such as those in the Maudlin scenario that can place it in a superposition of leftward and rightward momentum states. But since neither of these involve a matching "electron confirmation" for detection, the electron does not instantiate the Maudlin scenario. Thus, the Maudlin challenge lacks the required "slow-moving offer wave" and cannot be mounted at all. As we saw above, the experiment cannot be done with a photon OW, since it is not localizable and cannot be "preempted" by placement of an object in its path. The foregoing disposes of the Maudlin objection completely.

6.2 Interaction-Free Measurements

Elitzur and Vaidman (1993) pioneered the idea of "interaction-free measurements" (IFM) (see Figure 6.4). These experiments exhibit very clearly the counterintuitive, nonclassical nature of quantum events. We will see in this section that the transactional picture sheds new light on the nature of these experiments, and find that they are not really "interaction-free."

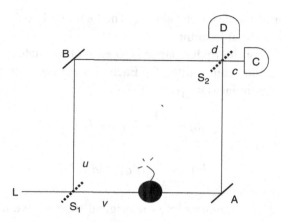

Figure 6.4 The Elitzur–Vaidman "bomb detection" experiment.

6.2.1 The Elitzur–Vaidman Bomb Detection Experiment

The original EV paper (1993) presents a way to examine a bomb to make sure it is working properly, but without activating (exploding) the bomb. Of course, the experiment is an idealization, but it provides an illustration of the way a quantum system can "probe" its environment without necessarily exchanging energy with it. The basic setup is illustrated schematically in Figure 6.4.

The laser L acts as a source of photons in a state we'll call $|s\rangle$. There are two beam splitters (half-silvered mirrors) S_1 and S_2, which transmit and reflect equal components of the incident state. Note that a photon described by a state such as $|v\rangle$ (corresponding to being found in arm v of the interferometer; see Figure 6.4) acquires a 90° phase change, corresponding to multiplication of a factor of i, upon reflection. (We disregard the total reflections at mirrors A and B because they don't affect the final result.) Thus, after two reflections, the state acquires a phase change of 180° and is multiplied by a factor of -1, and so on.

The interferometer is set up so that a photon entering the device can only be detected at detector C. Considering just the empty interferometer with no obstruction in either of the arms u or v, this is accomplished as follows. Let's call the initial photon state from the laser source $|s\rangle$. Upon passing through the first beam splitter S_1, its state is transformed as

$$|s\rangle \to \frac{1}{\sqrt{2}}(i|u\rangle + |v\rangle);\qquad(6.1)$$

that is, the initial state becomes a superposition of a transmitted component corresponding to arm v and a reflected component corresponding to arm u, with a phase shift factor of i as described above. (The factor of $1/\sqrt{2}$ indicates that these two components are equal in amplitude.)

Next, we have to consider what happens to each of the states $|u\rangle$ and $|v\rangle$ as they interact with the second beam splitter, S_2. Each of these states undergoes a splitting similar to that of the initial state $|s\rangle$, as follows:

$$|u\rangle \rightarrow \frac{1}{\sqrt{2}}(|c\rangle + i|d\rangle)$$

$$|v\rangle \rightarrow \frac{1}{\sqrt{2}}(i|c\rangle + |d\rangle). \qquad \text{(6.2a, b)}$$

If we substitute these expressions into the original state $|s\rangle$, we find that it evolves as follows:

$$|s\rangle \rightarrow \frac{1}{\sqrt{2}}(i|u\rangle + |v\rangle) \rightarrow \frac{1}{\sqrt{2}}\left[\frac{i}{\sqrt{2}}(|c\rangle + i|d\rangle) + \frac{1}{\sqrt{2}}(i|c\rangle + |d\rangle)\right]$$

$$= \frac{1}{2}[i|c\rangle - |d\rangle + i|c\rangle + |d\rangle] = i|c\rangle. \qquad \text{(6.3)}$$

Thus, destructive interference between components corresponding to path $|d\rangle$ prevents the photon from reaching detector D and that detector will never activate; photons will always be detected at C. Thus, detector D is called a "silent detector" in this type of experiment. In technical terms, the probability that the electron is on path c headed for detector C is given by the Born Rule, which prescribes that we square the projection of state (6.3) onto $|c\rangle$; thus we get

$$\text{Prob(C activated)} = -i \cdot i |\langle c|c\rangle|^2 = 1. \qquad \text{(6.4)}$$

Now, let's see what Elitzur and Vaidman have in mind as far as using this setup to examine a bomb without setting it off (see Figure 6.4). Keeping in mind the above analysis of the empty interferometer, consider the addition of an obstruction in arm v. We now have three possible experimental outcomes:

1. Detector C is activated.
2. Detector D is activated.
3. The photon is absorbed by the obstruction.

In this experiment, component $|v\rangle$ cannot reach S_2, so it cannot reach either detector. A photon described by $|v\rangle$ will inevitably be absorbed by the obstruction (outcome 3 above). The only component that has a chance of reaching the detector area is $|u\rangle$. Recalling that the original state $|s\rangle$ has equal components of $|u\rangle$ and $|v\rangle$ (5.1), the relevant probabilities are:

$$\text{Prob(C activated)} = \left| \langle c | \left(\frac{i}{\sqrt{2}} | u \rangle \right) \right|^2 = \frac{1}{4} \qquad (6.5a)$$

$$\text{Prob(D activated)} = \left| \langle d | \left(\frac{i}{\sqrt{2}} | u \rangle \right) \right|^2 = \frac{1}{4} \qquad (6.5b)$$

$$\text{Prob(photon absorbed by obstruction)} = \left| \langle v | \left(\frac{i}{\sqrt{2}} | v \rangle \right) \right|^2 = \frac{1}{2}. \qquad (6.5c)$$

If the photon is detected at D, then we know the bomb is active (in this idealization, counts as an obstruction) even though it has not been triggered.

6.2.2 A Quantum "Bomb"

Since the blocking object influences the ultimate nature of the photon detection even though the photon is not detected (absorbed) there, it can be thought of as a "silent detector." Hardy (1992b) provided a twist on the original Elitzur–Vaidman IFM. In his version, the bomb or other macroscopic "silent detector" is replaced by a quantum system: a spin one-half atom. The atom is prepared in a state of spin "up along x," which is then subject to a magnetic field gradient along the z direction and spatially separated so that it could be found in either of two boxes, one of which ("spin up along z," denoted by the state $|z \uparrow \rangle$) is placed in one path of the MZI. (Refer to Figure 6.5.) These "boxes" are assumed to be completely transparent to photons.

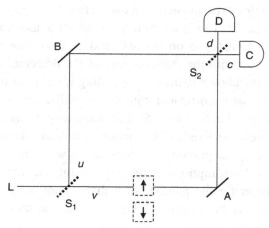

Figure 6.5 Hardy's version of the Elitzur–Vaidman interaction-free measurement with an atom replacing the bomb. *L* denotes a coherent (laser) photon source.

As noted in Hardy's discussion, the surprising feature of this experiment is that when detector D is activated, the atom must always be found in the box intersecting path v, in a well-defined spin state $|z \uparrow\rangle$; yet seemingly the photon did not interact with it, since the latter was detected at D and therefore was not absorbed by the atom. How is it possible for a photon which apparently went "nowhere near" an atom to dictate the state of the atom? Hardy's discussion is based on the idea of "empty waves," that is, Bohmian guiding waves in which the Bohmian particle is clearly absent yet the wave appears to have real effects. It is our purpose here not to address the Bohmian "empty wave" picture but to show that TI gives a natural and revealing account of this experiment.

The atom is understood to be in its ground state $|0\rangle$ unless otherwise specified. The atom's excited state – that is, its state when it has absorbed a photon – is denoted as $|1\rangle$. The state of the combined system of {photon, atom} starts out as:

$$|\Psi\rangle_i = |s\rangle \otimes \frac{1}{\sqrt{2}}(|z \uparrow\rangle + |z \downarrow\rangle) \tag{6.6}$$

where $|s\rangle$ denotes the photon source state. As before, the photon's state undergoes a phase shift of i upon reflection, so after passing through the first beam splitter S_1, the photon's state becomes $\left(\frac{1}{\sqrt{2}}\right)(i|u\rangle + |v\rangle)$. At this point the total system's state is

$$|\Psi\rangle_{S_1} = \frac{1}{2}(i|u\rangle + |v\rangle) \otimes (|z\uparrow\rangle + |z\downarrow\rangle)$$
$$= \frac{1}{2}(i|u\rangle|z\uparrow\rangle + |v\rangle|z\uparrow\rangle + i|u\rangle|z\downarrow\rangle + |v\rangle|z\downarrow\rangle). \tag{6.7}$$

The novel feature of this experiment is that one of the absorbers – the atom – is now entangled with the photon. This gives us three incipient transactions for the photon: (1) absorption by the atom, (2) absorption at C, and (3) absorption at D. Actualization of a photon transaction with the atom (or one of the detectors C or D) results in collapse not only of the photon to the corresponding state but also of the absorber to the state component that is correlated with that actualized photon transaction (we discuss this in more detail in Section 6.5). For macroscopic absorbers like detectors C and D that are well localized, this is inconsequential, but the atom's state is significantly affected by the photon detections, as we will see.

Under the idealized assumptions of the experiment, the atom in the state $|z \uparrow\rangle$ constitutes an absorber for the photon, such that a photon that was definitely in state $|v\rangle$ would definitely be absorbed.[7] Thus, the second term on the right-hand

---

[7] However, as discussed in Chapter 5, microscopic currents such as atoms really have only an amplitude to generate confirmations.

side of (6.7) serves to set up an incipient transaction corresponding to the photon being found on path v and thus being absorbed by the atom in state $|z \uparrow\rangle$. This will of course take the atom from its unexcited state $|z \uparrow; 0\rangle$ to its excited state $|z \uparrow; 1\rangle$.[8]

However, that incipient transaction may not be actualized, and there are other competing incipient transactions to consider. The photon OW component $|v\rangle$ correlated with the atomic state $|z \downarrow\rangle$ transforms into a superposition of states $|c\rangle$ and $|d\rangle$ corresponding to detectors C and D. Thus, the photon's entanglement with the atom creates a new possibility for the photon: the photon OW component $|v\rangle$ performs "double duty." This underscores the inappropriateness of trying to picture the photon OW as literally propagating "in spacetime"; it does not. Entities described by quantum states (i.e., kets and bracs) are not spacetime objects.[9] The photon OW component $|v\rangle$, by virtue of its entanglement with the atom, entertains two possibilities: (1) being absorbed by the atom in state $|z \uparrow\rangle$ or (2) missing the atom in state $|z \downarrow\rangle$ and being absorbed by either C or D. This situation illustrates the futility of clinging to a spacetime ontology for OW and CW, which are objects "too big" to fit into spacetime, as argued in Chapter 4. In quantitative terms, the entangled quantum state is characterized by six spatial dimensions (three for the photon and three for the atom) and, thus, if we are to take it as physically referring (i.e., if we are being realist about quantum theory), the entity to which it refers simply cannot exist in three-dimensional space. The state of the photon and the atom is nonseparable (i.e., cannot be written as a product state), and this means we cannot separate their spatial coordinates and treat them each as pursuing an independent trajectory. They do not.

So, we must acknowledge that *OW and CW do not pursue simple spacetime trajectories*. They are of course constrained by aspects of the experimental arrangement, but those constraints likewise involve quantum possibilities that govern the kinds of transactions that can occur. That is, it must always be kept in mind that the experimental apparatus is fundamentally composed of quantum systems that exist and interact on the quantum (pre-spacetime) level, and that the phenomena we see and experience of the apparatus are just the results of transactions.

---

[8] Note that the collapse, not only of an OW to one particular component but also of microscopic absorbers in superpositions, can be seen as the way in which events can be actualized in a true "becoming" picture of spacetime. That is, absorbers are pre-spatiotemporal possibilities and as such serve as creators/facilitators of determinate events through non-unitary reductions. It is events and their connections (actualized photons), not rest-mass quantum systems, that are elements of spacetime. I return to the issue of spacetime "becoming" in Chapter 8.

[9] Recall that only actualized emission and absorption events, and the transferred photons between these, are spacetime objects. Transferred photons are represented by projection operators, not kets or bracs.

Besides the incipient transaction for the photon's absorption by the atom, the remaining incipient transactions are based on the combined state:

$$|\Psi\rangle_f = \frac{1}{2\sqrt{2}}|d\rangle|z\!\uparrow\rangle + \frac{i}{2\sqrt{2}}|c\rangle|z\!\uparrow\rangle + \frac{i}{\sqrt{2}}|c\rangle|z\!\downarrow\rangle. \tag{6.8}$$

Let us first focus our attention on the "silent detector" case, detection at D, represented by the first term in (6.8). The amplitude for this component applies to the two correlated degrees of freedom, that is, the atom and the photon. To be fully precise going forward, it must be emphasized that (as noted in Section 6.1) the atom is an absorber with nonvanishing rest mass and does not have its own confirmations; it is subject to Schrödinger evolution in the Stern-Gerlach device and only undergoes non-unitary collapse or reduction via photon transactions (by way of emissions, absorptions, or, as in the case at hand, through entanglement with photons). The incipient transaction corresponding to the first term, $(1/(2\sqrt{2}))|d\rangle|z\!\uparrow\rangle$, is set up by photon OW and CW to/from the D detector, each with an amplitude of $1/(2\sqrt{2})$.[10] The square gives us the Born probability of 1/8 for actualization of this transaction.

We see from the combined state (6.8) that a D transaction can only occur for an atom in the "blocking" state $\langle z\!\uparrow; 0|$ (corresponding to the photon being in arm *u*), which explains why the atom's initial superposition must be "collapsed" whenever the photon is detected at D. Specifically, one finds that the photon components corresponding to D for the state $|z\!\downarrow\rangle$ mutually cancel. This atomic state effectively returns the interferometer to its original function, as in (6.3), in which there is nothing blocking arm *v*.

For completeness, let us also consider the incipient transaction for the photon's detection at C. This arises from the last two terms in (6.8), which we will refer to as $|\Psi_C\rangle$ (note that this is a truncated state with overall amplitude less than unity):

$$|\Psi_C\rangle = |c\rangle\left(\frac{i}{2\sqrt{2}}|z\!\uparrow\rangle + \frac{i}{\sqrt{2}}|z\!\downarrow\rangle\right). \tag{6.9}$$

The normalized atomic state corresponding to the photon component $|c\rangle$ is the relative state $|a\rangle$ (we discuss the concept of relative states in more detail in Section 6.5):

$$|a\rangle = i\left(\frac{1}{\sqrt{5}}|z\!\uparrow\rangle + \frac{2}{\sqrt{5}}|z\!\downarrow\rangle\right). \tag{6.10}$$

---

[10] For specifics on why the photon OW is characterized by this amplitude, see Section 6.5.

$|a\rangle$ is simply the state of the atom corresponding to a photon detection at C. However, as noted, the amplitude of the atomic state $|a\rangle$ present in the two-quantum state component $|\Psi_C\rangle$ is less than unity, specifically:

$$|\Psi_C\rangle = |c\rangle\left(\frac{i}{2\sqrt{2}}|z\uparrow\rangle + \frac{i}{\sqrt{2}}|z\downarrow\rangle\right) = \frac{\sqrt{5}}{2\sqrt{2}}|c\rangle|a\rangle$$

$$= \frac{\sqrt{5}}{2\sqrt{2}}|c\rangle\left(\frac{i}{\sqrt{5}}|z\uparrow\rangle + \frac{i}{\sqrt{5}}|z\downarrow\rangle\right). \tag{6.11}$$

This means that the photon OW reaching detector C has a corresponding amplitude of $\sqrt{5}/(2\sqrt{2})$. It will generate a CW of matching amplitude, leading to a Born probability of 5/8 for this transaction. If this transaction is actualized, the atom will be collapsed in the state $|a\rangle$, an indefinite spin state with respect to both Z and X. One can of course do further Stern–Gerlach measurements on the atom to check that it is in this state.

The above situation only seems paradoxical if we insist on thinking of quantum objects as classical corpuscles carrying energy and momentum along specific trajectories. This "billiard ball" notion is what is denied in TI: quanta are not corpuscles pursuing trajectories. Amplitudes describe offer and confirmation waves which themselves do not transfer energy, but which can give rise to transactions. It is only the completed (actualized) transactions that transfer energy and other conserved quantities, and which therefore activate detectors.

6.3 Delayed Choice Experiment

John Wheeler's delayed choice experiment (DCE) is an example of what appears to be a quantum "temporal paradox." In this section, we'll apply the transactional picture to the DCE.

The standard (non-TI) presentation of the DCE is as follows (see Figure 6.6). (1) At $t = 0$, a photon is emitted toward a barrier with two slits A and B. (2) At $t = 1$, the photon passes the barrier. (3) The photon continues on to a screen S on which one would expect to record (at $t = 2$) an interference pattern as individual photon detections accumulate. (4) However, the screen may be removed before the photon arrives (but after it has passed the slit barrier), revealing two telescopes focused on each slit. (5) If this happens, the two telescopes T will perform a "which slit" measurement at $t = 3$, and the photon will be detected at one or the other telescope, indicating that it's in a "which slit" state (i.e., there is no interference). The decision as to whether to remove S or not is made randomly by the experimenter.

According to this standard account, the photon has already passed the plane of the slits before the observer has decided whether to measure "which slit" or not.

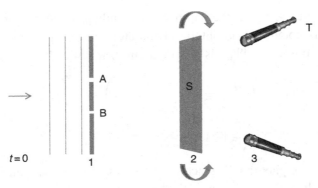

Figure 6.6 The delayed choice experiment.

Thus, at a time $1 < t < 2$ prior to the observer's choice, there is apparently no clear "fact of the matter" about the photon's state, including whether or not it has "interfered with itself."[11] Wheeler famously interpreted his experiment as demonstrating that "no phenomenon is a real phenomenon until it is an observed phenomenon."[12] While many have interpreted Wheeler's comments about the DCE as endorsements of the fundamentality of consciousness, he actually seemed to prioritize physical accounts of measurement interactions, as evidenced by these comments:

It from bit. Otherwise put, every *it* – every particle, every field of force, even the space-time continuum itself – derives its function, its meaning, its very existence entirely – even if in some contexts indirectly – from the apparatus-elicited answers to yes-or-no questions, binary choices, bits. It from bit symbolizes the idea that every item of the physical world has at bottom – a very deep bottom, in most instances – an immaterial source and explanation; that which we call reality arises in the last analysis from the posing of yes-no questions and the registering of equipment-evoked responses; in short, that all things physical are information-theoretic in origin and that this is a *participatory universe.*

(Wheeler, 1990)

This take is actually very much in harmony with the transactional picture, since TI gives a specific, physical account of what Wheeler calls "the apparatus-elicited answer" and "the registering of equipment-evoked responses." Of course, where traditional quantum theory could not provide any specifics of what yields such "answers," TI does so, and locates the fundamental participatory aspect in the

---

[11] I should note that David Ellerman argues against any retrocausation associated with the DCE, saying that the photon is always in a both-slits state and is just projected into a which-slit state at the final measurement if that is what is performed (Ellerman, 2015). However, equating the photon solely with its prepared quantum state, a ket, assigns it the wrong units (square root of energy) and therefore cannot explain why the detected photon delivers real energy to the detector. TI thus improves upon this account, since its incipient transactions have the correct energy units. Nevertheless, in principle, the DCE can be done with a massive particle, which would indeed be in the both-slits state until its detection (which would still involve a photon transaction).

[12] Wheeler (1990).

non-unitary interaction between emitters and absorbers that is available only from the direct-action theory of fields. That is, we need an interaction between emitters and absorbers to give rise to an "it" – really, a transferred photon. The "bit," according to RTI, corresponds to the quantum substratum – the *potentiae* that are emitters and absorbers and their interactions (by way of quantum-level fields). The latter is what Wheeler calls an "immaterial source." He views the "bit" source as "immaterial" because, as is traditional among physicists, he views only spacetime phenomena as "material" or "physical." In the final analysis, such ontologically flavored terms are just semantic labels; the key point is that one must allow that reality, as describable by physical theory, comprises a vast amount of content beyond the spacetime manifold.

Now let us consider some specifics of the TI account of the DCE. First, remember that the transferred photon is always the result of a NU-interaction; there must be absorber(s) available for any emitter in order to create a photon. The transferred photon is represented by a projection operator. This means that the photon is not emitted *until* there are confirmations, a set of incipient transactions created, and one is actualized. The photon is the result of the actualized transaction. So, it is *not* the case that "a photon is emitted, passes the slits, and is then absorbed." Rather, there are interactions among the quantum *potentiae* (pre-spatiotemporal entities) that are the emitters and absorbers, such that the OW and CW *together* set up incipient transactions corresponding to various types of real photons (i.e., momentum, polarization) and one of these is actualized. The spacetime interval corresponding to the transferred photon does not exist until there is an actualized transaction (we consider the emergence of spacetime in more detail in Chapter 8). All the other interactions (i.e., everything except that actualized transferred photon) take place in the quantum substratum (mathematically described by Hilbert space, not 3+1 spacetime). We see that this account dovetails with Wheeler's view that

every *it* – every particle, every field of force, even the space-time continuum itself – derives its function, its meaning, its very existence entirely ... from the apparatus-elicited answers to yes-or-no questions, binary choices, bit.[13]

With regard specifically to the delayed aspect: the OW/CW NU-interaction is constrained by relativity's required finite time of passage (with respect to any inertial frame) of the real photon from the emitter to the receiving ("winning") absorber, that is, electromagnetic energy transfer must respect the finite speed limit of light. This is an example of the manner in which relativity and quantum theory are harmonious: relativity places constraints on the manner in which elements of spacetime (events and their connections) emerge from the quantum level, and

---

[13] However, in Chapter 8 we'll see that in RTI the emergent spacetime manifold is not a continuum.

quantum theory precisely accommodates those constraints. The upshot is that no CW can be generated from absorbers whose CW generation would correspond to a spacelike interval from the OW emission. OW and CW can only be generated if the OW/CW generations are separated by a null interval. Thus, the RTI picture of the DCE is that the photon source, together with the two-slit screen, generates an OW in a "both slits" state, that is,

$$|\Psi\rangle = \left(\frac{1}{\sqrt{2}}\right)[|A\rangle + |B\rangle]. \tag{6.12}$$

If the detection screen is left in place (establishing a null interval for a photon transfer), each individual absorber in the screen (corresponding to some position x) will receive a both-slits OW component:

$$\langle x|\Psi\rangle|x\rangle = \left(\frac{1}{\sqrt{2}}\right)[\langle x|A\rangle + \langle x|B\rangle]|x\rangle \tag{6.13}$$

and will generate a corresponding both-slits CW with the complex conjugate amplitude:

$$\langle\Psi|x\rangle\langle x| = \left(\frac{1}{\sqrt{2}}\right)[\langle A|x\rangle + \langle B|x\rangle]\langle x|. \tag{6.14}$$

The outer product of the OW and CW components yields for each absorber x a projection operator representing the incipient transaction:

$$\langle x|\psi\rangle\langle\psi|x\rangle|x\rangle\langle x| = \frac{1}{2}((\langle x|A\rangle + \langle x|B\rangle)(\langle A|x\rangle + \langle B|x\rangle))|x\rangle\langle x|$$

$$= \frac{1}{2}|(\langle x|A\rangle + \langle x|B\rangle)|^2|x\rangle\langle x|. \tag{6.15}$$

The square of a sum of amplitudes shows the usual interference for a "both-slits" experiment. Of course, we get a whole set of these corresponding to all values of x on the screen.

As discussed in Chapter 3, the sum of all the weighted projection operators corresponding to the set of incipient transactions for the various values of x is precisely the mixed state in von Neumann's non-unitary "Process 1" or measurement transition.

If the detection screen is removed, there is no longer a null interval between the OW generation and a CW generation at the screen, so the CW generation instead takes place at the more distant telescopes, which receive the single-slit or "which way" components

$$|\Psi_A\rangle = \frac{|A\rangle}{\sqrt{2}}, \qquad |\Psi_B\rangle = \frac{|B\rangle}{\sqrt{2}} \tag{6.16}$$

and correspondingly generate the CW components

$$\langle\Psi_A| = \frac{\langle A|}{\sqrt{2}}, \qquad \langle\Psi_B| = \frac{\langle B|}{\sqrt{2}} \qquad (6.17)$$

which lead to the set of two incipient transactions:

$$|\langle A|\Psi\rangle^2||A\rangle\langle A| = \frac{1}{2}|A\rangle\langle A| \qquad (6.18a)$$

$$|\langle B|\Psi\rangle^2||B\rangle\langle B| = \frac{1}{2}|B\rangle\langle B|, \qquad (6.18b)$$

leading to a detection distribution with no interference, but instead a sum of squared amplitudes.

In each case, the generated OW has the same form, that is, a two-slit state. What changes based on the choice is the set of CWs generated. As we see from the above, the set of CWs dictate the projection operators that define the measurement basis (this was also discussed in Chapter 3). Thus it is the CW (not the OW) that really dictates what kind of photon will be transferred. In this sense, as Wheeler noted, it is a measurement interaction in the present that acts to "bring about the past," where in this case the past is the emission event and the transferred photon. In other words, no real photon even exists *unless and until* there is an appropriate absorber available. The real photon's emission event and its passage are brought into being retroactively, in that sense. But it must be kept in mind that the real photon is *not* the OW or CW independently, so the delayed choice does not "overwrite" the career of a real photon already in progress to a detector. The photon is the result of the final, actualized transaction, represented by a projection operator weighted by the appropriate Born probability.[14] Again, the interactions leading up to the creation of the real photon (represented by the "winning" projection operator) take place in the quantum substratum and are not spacetime processes. Nevertheless, obviously they are very real and physically consequential.

Thus, RTI understands quantum theory as instructing us that we must expand our view of what is "physically real" to include the quantum substratum described by Hilbert space, where that substratum consists of *real possibilities* (*res potentiae*) that are necessary (but not sufficient) precursors to spacetime events.[15] In taking this step, we must let go of the usual assumption that physical processes take place against a spacetime background or within a "spacetime container."

---

[14] It's also important to note that when dealing with single photons (Fock states, having precise energies), there is no way to ascertain their exact "time of emission" and "time of arrival." So it's misleading to describe the experiment as being able to pin down exact times for these events.

[15] Cf. Kastner et al. (2018). The *potentiae* are not sufficient in that there is reduction to only one of a set of incipient transactions; not all result in spacetime processes such as photon transfer.

We also need to recognize the usual quantum state or ket as only part of the story, since the ket alone (as obtained from the quantum electromagnetic field acting on the vacuum state) has units of the square root of energy (per wave vector) and therefore cannot account for the photon's transfer of energy and other conserved quantities to the detector (see Section 5.2.5, discussion under Equation (5.23)). The transfer of conserved physical quantities is properly represented by the outer product of the OW and CW, that is, a projection operator characterized by the correct physical units. RTI provides a physical account of the applicability of the projection operators, which reflect properties of both the source and the detector (absorber): both participate in the creation of the quantum of momentum/ energy/angular momentum that is the real photon ultimately transferred between the source and the receiving detector as a result of the actualized transaction. This allows a full ontological account of the photon's transfer that reflects the influence of both the source and the absorber configuration, as discussed in this section. The projection operators are multiplied by the square of the overlap of the preparation state with the detection state, which is naturally interpreted as the Born probability that the detector will receive the full amount of conserved quantities represented by the projection operator.

6.4 Quantum Eraser Experiments

The term "quantum eraser experiment" refers to a class of experiments involving a pair of correlated photons. One of the pair is termed the "signal" photon and the other is termed the "idler" photon. The signal photons are directed into a two-slit apparatus and, depending on what is done with their paired idler photons, an interference pattern may or may not be seen for the appropriate subset of the signal photons. Some versions of the experiment send a single photon through the two-slit apparatus and then convert it into two correlated photons after the two slits; this is the version discussed below. The term "erasure" is actually a misnomer, because no information is really erased, as we will see below. However, it refers to the process in which a particular kind of measurement of the idler photon makes unavailable any "which slit" information that might be associated with the signal photon. There are separate detection arrangements for the signal and idler photons, and their separate detection information is sent to a coincidence counter to keep track of the pairs.

6.4.1 Details of a Quantum Eraser Experiment

The signal photons in this type of experiment are always detected at a detector S which is scanned across positions x to determine the count at each position (refer

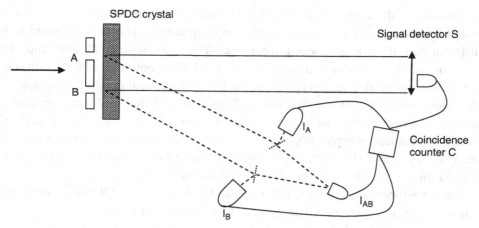

Figure 6.7 A quantum eraser experiment. (I am indebted to Ross Rhodes for suggestions for this and Figure 6.8.)

to Figure 6.7). That information is sent to the coincidence counter. The idler photons may be subjected to (1) a "which slit" measurement or (2) a "both slits" measurement, with two detectors corresponding to the possible outcomes of (1) and (2). Idler detections are sent to the coincidence counter, which keeps track of the correlated pairs.

Any idler photon that activates detectors (1) is correlated with a signal photon with "which slit" information, and any idler photon that activates detectors (2) reflects a superposition of slit states. It is only by looking at appropriate coincidences that one can see the above effects.

Recall from Chapter 3 that one can do a two-slit experiment with one particle (in that case, an electron) along with an auxiliary measurement by another particle (in that case, a photon). In that example, the electron played the part of the "signal photon" and the photon played the part of the "idler." What the experimenter chooses to do with the photon (i.e., how sharp a measurement to make) affects whether or not electron self-interference takes place (i.e., whether or not one sees an interference pattern for the electrons or a distribution corresponding to "which slit" paths). Quantum eraser experiments extend that basic setup by replacing the choice of how sharp a measurement to make with a choice of what kind of process is imposed on the idler.

In the usual approach to discussing these types of experiments, it is assumed that the signal photon either "went through a particular slit" or "went through both slits," depending on the kind of measurement performed on the idler photon. This seems to imply the very mysterious idea that what is done with the idler photon can materially affect the signal photon's spacetime trajectory. However, in TI, the influences involved are not at the level of "we poked one photon and somehow ended up instantly (or even retroactively!) poking another photon in a completely different part of the experimental apparatus." This is because in TI (and arguably even in standard quantum theory; see note 5) the photon is not a corpuscle pursuing a spacetime trajectory. Rather, the OW is a physical possibility created by the source together with the two-slit configuration, and that OW has a particular state – in this case, the two-slit (two-photon) state, irrespective of what kind of final measurement is made. So, in all these variations on the two-slit experiment, the OW is a "both slit" entity. The possible transactions available to that entity depend on the absorber configuration that generates CWs.[16]

The experimental setup of the version by Kim et al. (2000) is depicted schematically in Figure 6.7. The original OW, which can be written as

$$|\Psi\rangle = \left(\frac{1}{\sqrt{2}}\right)[|A\rangle + |B\rangle], \tag{6.19}$$

is converted into a two-photon correlated OW by way of a *spontaneous parametric down conversion* (SPDC) process. This process duplicates each "which slit" component but with opposite polarizations for each of the two photons. If we don't explicitly write the polarization states (which serve to correlate the two photons and enable experimenters to send them off into different directions), the two-photon state can be written as

$$|\Psi\rangle = \left(\frac{1}{\sqrt{2}}\right)[|A\rangle_S|A\rangle_I + |B\rangle_S|B\rangle_I] \tag{6.20}$$

where the first and second kets in each term correspond to the signal and idler, respectively. The signal photon OW components are sent to detector S and the idler photon OW components to another detector assembly, which is actually a system of beam splitters and mirrors with four subdetectors I_A, I_B, I_{AB}, and I_{BA}. (I_{BA} is not shown in the diagram for simplicity.) The latter two both detect interference patterns; they are just exactly out of phase. They detect "fringe" and "antifringe" patterns corresponding to the states

$$|AB\rangle = \frac{1}{\sqrt{2}}(|A\rangle + i|B\rangle)$$

$$\tag{6.21}$$

$$|BA\rangle = \frac{1}{\sqrt{2}}(|A\rangle - i|B\rangle).$$

It's important to note that the sum of the two interference patterns corresponding to these states is the same as the sum of the patterns for the states $|A\rangle$ and $|B\rangle$.

---

[16] The phrase "absorber configuration" here includes all components of the experiment including the relevant Hamiltonian.

Figure 6.7 schematically shows the idler detectors I_A, I_B, I_{AB}. The beams corresponding to passage through A and B are split by half-silvered mirrors. The reflected component of each is sent to detectors I_A and I_B, respectively, which provides a "which path" measurement of its signal photon partner (just as in the use of telescopes aimed at each slit in the two-slit experiment), and the transmitted components of each will be recombined and may reach the other two detectors I_{AB} and I_{BA}. The recombined A and B beam components detected by I_{AB} and I_{AB}' can no longer provide "which path" information, and this is the origin of the term "quantum eraser."

Meanwhile, the signal photon heads toward the movable detector S, which is located at varying positions x for different runs of the experiment. If the signal photon is detected at position x, detector S sends a count to the coincidence counter. The idler detection for that run, wherever it occurs, is matched via the coincidence counter to its partner signal photon. (If the signal photon is not detected at x, it cannot be matched to its idler partner and that run does not show up in the coincidence count, so there is no data for that run.) In this way, the experimenters have a joint count; that is, for all signal photon detections at position x, they can see how many idler photons were detected at each of the idler detectors. Those signal photons whose idlers were detected at I_A and I_B turn out (as predicted by standard quantum mechanical calculations of the relevant probabilities) to be distributed in a non-interfering "single-slit" distribution, while those whose idlers were detected at I_{AB} form an interference pattern. The latter situation gave rise to the idea that the "which slit" information has been "erased," although that is misleading since neither photon of a correlated pair starts out with any particular kind of information at an individual level, since they are in improper mixed states. This issue is discussed in Kastner (2019b).

6.4.2 The TI Account

Let us now consider the TI account of this experiment. The total system's OW is as given in (6.20). Detector S generates a CW component $\langle x|$ corresponding to its position in any given run, and the signal photon may therefore be absorbed at $S(x)$, where this notation specifies the position of S for any given run. However, neither photon has an independently well-defined OW, since they are components of an entangled two-photon state; the OW is well defined only for the two-photon state, and all CW responses are really two-photon CWs.[17] The idler component is locally

---

[17] It's important to keep in mind that photons, unlike quanta with rest mass, experience no passage of time. For photons pairs created by down-conversion, there is no sense in which one set of the entangled photons' CW are generated "before" the other's, regardless of the separation of the detectors themselves (whether timelike or spacelike). The CWs are all two-photon CWs and are not locally separable.

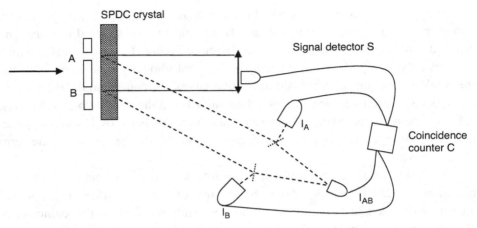

Figure 6.8 "Delayed quantum erasure." As argued in Kastner (2019b), in fact nothing is delayed, and no information is erased.

divided by the beam splitters to introduce two different measurement bases: the "which way" basis and the "both ways" basis.

Before looking at the quantitative details, we note that the experiment may also be implemented with a "delayed" aspect (see Figure 6.8): the idler photon detection may be delayed until after the signal photon has been detected at S. This makes the experiment seem astounding from the standard point of view. A typical discussion of a variation of the delayed version says, in part:

Before photon *p* [the idler photon] can encounter the [erasing] polarizer, *s* [the signal photon] will be detected. Yet it is found that the interference pattern is still restored. It seems *s* knows the "which-way" marker has been erased and that the interference behavior should be present again, without a secret signal from *p*. How is this happening? It wouldn't make sense that *p* could know about the polarizer before it got there. It can't "sense" the polarizer's presence far away from it, and send photon *s* a secret signal to let *s* know about it. Or can it? And if photon *p* is sensing things from far away, we shouldn't assume that photon *s* isn't.[18]

The above discussion includes the usual metaphysical assumption that I believe needs to be rejected, that is., that an emitted photon is pursuing a spacetime trajectory. This makes the phenomena seem particularly bizarre, necessarily involving remote sensing and/or foreknowledge on the part of photons considered as material corpuscles. Meanwhile, the TI account of the delayed choice version of this experiment simply involves a set of two-photon incipient transactions; their time order is completely unimportant. To see this, let us first take into account that

---

[18] Excerpted from Orozco (2002).

the signal photon's absorber responses come from the movable projector $S(x)$. For a particular value of x, the total two-photon state is:

$$|\Psi\rangle_x = \left(\frac{1}{\sqrt{2}}\right)[\langle x|A_S\rangle|x\rangle|A_I\rangle + \langle x|B_S\rangle|x\rangle|B_I\rangle]. \qquad (6.22)$$

For each value of $S(x)$, we will get four different two-photon OW components based on the four local separations of the idler OW component. The two reflected idler components reaching the detectors I_A and I_B are attenuated by a factor of $i/\sqrt{2}$. The other two components are transmitted through the initial beam splitters, being attenuated by a factor of $1/\sqrt{2}$, and enter an interferometer setup equipped with detectors I_{AB} and I_{BA}. In the interferometer, the components A and B transform as follows in terms of the detection states $|AB\rangle$, $|BA\rangle$:

$$|A\rangle \rightarrow \frac{1}{\sqrt{2}}(i|AB\rangle + |BA\rangle)$$

$$|B\rangle \rightarrow \frac{1}{\sqrt{2}}(|AB\rangle + i|BA\rangle). \qquad (6.23)$$

For convenience, let us abbreviate the amplitude $\langle x|A_S\rangle$ as $A(x)$, and similarly for B. The two-photon OW component corresponding to the interferometer section evolves as follows:

$$|\Psi\rangle_x = \left(\frac{1}{2}\right)|x\rangle[A(x)|A_I\rangle + B(x)|B_I\rangle]$$

$$\rightarrow \left(\frac{1}{2\sqrt{2}}\right)|x\rangle[A(x)(i|AB\rangle + |BA\rangle) + B(x)(|AB\rangle + i|BA\rangle)]$$

$$= \left(\frac{1}{2\sqrt{2}}\right)|x\rangle[(iA(x) + B(x))|AB\rangle + (A(x) + iB(x))|BA\rangle]. \qquad (6.24)$$

Thus, the four two-photon OW components reaching each pair of detectors are

Detectors $S(x)$, I_A: $\frac{i}{2}A_S(x)|x\rangle|A_I\rangle$

Detectors $S(x)$, I_B: $\frac{i}{2}B_S(x)|x\rangle|B_I\rangle$

Detectors $S(x)$, I_{AB}: $\frac{1}{2\sqrt{2}}(iA_S(x) + B(x))|x\rangle|AB\rangle$

Detectors $S(x)$, I_{BA}: $\frac{1}{2\sqrt{2}}(A_S(x) + iB(x))|x\rangle|BA\rangle$.

Each pair of detectors generates a corresponding CW, yielding an absolute square of the amplitudes as discussed in Chapters 3 and 4. For detections at I_{AB} and I_{BA}, which involve the squares of sums of amplitudes, we obtain opposite (exactly out of phase) interference patterns based on the varying values of x. The crucial point is that these patterns obtain regardless of the time order of the photon detections; they are insensitive to any time order. The signal photon detections at $S(x)$ and the idler photon detections at their four detectors are mutually interdependent, since they arise from the above two-photon transactions.

Of course, one can also give an account from within standard quantum theory that shows that there is no real "delayed" effect and no "erasure," since prior detections of the signal photon at x can be seen as conditionalizing the probabilities of detection of the idler photon at its respective detectors, thus enforcing the correlations. Readers interested in a more detailed debunking of the "erasure" claims of this experiment may consult Kastner (2019b). Fearn (2016) also provides a transactional account of the QE experiment that presents more details of the interference patterns obtained.[19] In particular, there is no real "delayed choice" in the sense of Wheeler's DCE discussed previously, since each signal photon always undergoes a both-slits measurement. The idler is then projected into a state corresponding to that of its signal partner's "both slits" detection, and correlations are seen only through coincidence counts.

Finally, we should emphasize that the analysis of entangled photons, as in the above example, differs from that of a photon entangled with a rest-mass objects such as an electron or atom, since such fermionic systems are emitters and absorbers and don't have their own OW and CW. The latter situation is discussed in the next section. Even though the specifics of the transactional process are different in the two cases, we will see that emitters and absorbers undergo "collapse" or reduction in a manner governed by exactly the same probabilities. This is why, at an earlier stage of the development of TI, it was adequate to apply an OW/CW analysis to such systems, even though technically they do not possess their own OW/CW.

6.5 Transactions and Decoherence

The term "decoherence" describes the loss of a quantum system's ability to exhibit interference effects such as the fringes in a two-slit experiment.[20] In more technical terms, one can find evidence of decoherence when the system's density matrix is diagonal with respect to the observable being measured, that is, when its off-diagonal elements vanish. The density matrix of the system is obtained by "tracing over" the degrees of freedom of the measuring apparatus to obtain a "reduced density matrix" for the system. The diagonal basis of the reduced density matrix is commonly referred to as the "pointer basis," where the pointer is some macroscopic system with different possible readings, or a set of macroscopic

---

[19] However, one correction is in order for Fearn's account. She takes the inner product of the entire idler OW and CW expressed as a vector and dual vector, respectively, and then appears to assume that the idler basis vectors are not orthogonal in order to retain the interference. However, the interaction of the full OW and CW is represented by an outer product (i.e., it is a projection operator, not an inner product). In the "which way" basis, what we then get is a density matrix with nonzero off-diagonal terms, representing interference (see Section 6.5).

[20] This section is based on material in Kastner (2020a).

phenomena such as detector clicks, corresponding to the outcomes of the observable being measured.

It should be noted that the decoherence program presupposes that measurement involves only the establishment of correlations in a unitarily evolving entanglement of two degrees of freedom (such as a system and measuring apparatus). Among the main pioneers of the study of decoherence are Joos, Zeh, Zurek, Omnès, and others.[21] This is a rigorous and well-elaborated research program that has received a large quantity of experimental corroboration, and thus has clear empirical validity. However, this conventional unitary-only decoherence program runs into interpretive problems, since the system's reduced density matrix is an improper mixture; it cannot be interpreted as describing a situation in which the system really has a definite outcome (as discussed in Section 1.3.1). This concern about improper mixtures is explained in a pedagogically clear manner by R. I. G. Hughes (1989, section 9.6). In short, the fact that one has eliminated interference effects by tracing over the pointer degrees of freedom does not mean that one has eliminated the superposition of two contrary eigenstates on the part of the measured system. Thus, the reduced density matrix in a unitary-only account does not license a conclusion that the system is "really" in an eigenstate corresponding to the observed outcome. This means that the inconsistency between theory and empirical results, that is, the measurement problem, remains unsolved under the conventional decoherence program, and it can be no more than a "for all practical purposes" (FAPP) procedure.

In addition to the fact that decoherence does not solve the measurement problem, the program's initial goal of explaining the natural emergence of seemingly definite measurement outcomes in an apparently "classical" phenomenal world has arguably not been met. This program, known as "quantum Darwinism," fails due to circularity (e.g., Zanardi, 2001; Dugić and Jeknić-Dugić, 2012; Kastner, 2014b).[22] While we don't address this issue in detail here, the basic problem is that the program of showing classical emergence based on decoherence needs to help itself to essentially classical or proto-classical initial conditions, such

---

[21] I do not attempt to do full justice here to the extensive history and literature of the decoherence program. Among important pioneering works are Joos and Zeh (1985), Omnès (1997), and Zurek (2003). Additional relevant references can be found in Kiefer and Joos (1999).

[22] Zurek attempts to address this issue in terms of an argument of transcendental form. He remarks: "As the interpretation problem does not arise in quantum theory unless interacting systems exist, we shall also feel free to assume that an environment exists when looking for a resolution" (Zurek, 2003). However, under the unitary-only assumption, the fact that quantum systems are interacting need not lead to classically recognizable pointer states. In order for a classically recognizable preferred pointer basis to emerge under the unitary-only restriction, one must invoke special assumptions. These take the form of a computational basis reflecting classically recognizable information, initial separability of appropriately distinct degrees of freedom, and/or the assignment of many more degrees of freedom to the environment or apparatus than to the designated system. Such assumptions essentially incorporate classicality at the outset; hence, the program is circular.

as appropriate separability of the universal Hilbert space, or a "logical basis" preferring classically localized phenomena, that have no independent justification.

In what follows, we will recall the essentials of the conventional unitary-only account of decoherence. We will then consider how TI fills in some crucial missing steps that are needed in order to gain a proper mixture for the system that can be legitimately interpreted as reflecting the determinacy of one outcome. That is, the statistical description of the system's reduced density matrix is then legitimately based on ignorance of the actualized outcome.

6.5.1 Decoherence: Two Aspects

First, we need to disambiguate two physically distinct aspects to decoherence: (1) the resolution of the induced measurement correlation designed to measure a particular observable and (2) the rate of repetition of the relevant physical interaction. Aspect (1) describes the magnitudes of the system's off-diagonal reduced density matrix elements due to a single correlating interaction. These are governed by the inner product of the pointer states $\langle \varphi_m, \varphi_n \rangle$, also called the "decoherence function." Meanwhile, (2) reflects the rate of decrease of the off-diagonal elements due to repeated measurement interactions. A common example of a correlating interaction resulting in various degrees of decoherence is photon emission. Kokorowski et al. (2000) present an instructive account of this process. They discuss an experiment involving atoms injected into an interferometer and subject to stimulating radiation. This causes them to emit photons in one of two possible "pointer" states correlated with each path of the interferometer, allowing for a "which way" measurement.

Let us now consider in more detail (1): the resolution of the basic measurement interaction, characterized by the decoherence function.[23] A correlation for purposes of measurement is induced by an interaction Hamiltonian that effectively entangles the system under study and a measurement apparatus. In such an interaction, each system state $|n\rangle$, corresponding to an eigenvalue n of the observable being measured, is correlated to a pointer state of the apparatus, here designated $|\varphi_n\rangle$. It's important to note that these correlated pointer states need not be mutually orthogonal. Following Bub's presentation, let us call these possibly non-orthogonal pointer states "relative pointer states," since they are defined relative to each of the system observable's eigenvalues n.

In contrast, when we find an outcome on our measuring apparatus, say, x, these outcomes always correspond to an orthogonal pointer basis $\{x\}$. So it's important to note that in general, we don't directly detect the relative pointer states $|\varphi_n\rangle$.

---

[23] This section follows the pedagogical treatment of Kiefer and Joos (1999).

Rather, we detect outcomes in the pointer basis. The only case in which we directly detect the relative pointer states is when they coincide with the pointer basis. This constitutes a maximally sharp or ideal measurement. In this case, the decoherence function $\langle \varphi_m, \varphi_n \rangle = \langle x_m, x_n \rangle = 0$. In an ideal measurement, the system's reduced density matrix is diagonal with respect to the pointer basis. But in general, a measurement interaction need not be sharp or ideal, so that in general the decoherence function is nonvanishing, and the off-diagonal elements remain finite.

Suppose the quantum system is initially in some arbitrary state, $|\psi\rangle = \sum_n c_n |n\rangle$, and the apparatus in an initial ready state $|\varphi_0\rangle$. The evolution of the combined total system looks like

$$|\psi\rangle |\varphi_0\rangle \rightarrow \sum_n c_n |n\rangle |\varphi_n\rangle. \tag{6.25}$$

We obtain the density matrix for the total system by forming the projection operator corresponding to the state in the right-hand side of (6.25). To get the system's reduced density matrix, we trace over a pointer basis and find:

$$\rho_S = \sum_{n,m} c_m^* c_n \langle \varphi_m | \varphi_n \rangle |m\rangle \langle n|. \tag{6.26}$$

The sizes of the off-diagonal elements in (6.26) are governed by the inner products of the relative pointer states, $\langle \varphi_m | \varphi_n \rangle$, that is, the decoherence function. If $\langle \varphi_m | \varphi_n \rangle \approx 0$, the system's reduced density matrix is approximately diagonal. This corresponds to a "sharp" or "strong" measurement, that is, one in which the correlation provides good resolution for the system observable. However, we can also have cases in which the decoherence function, and therefore the off-diagonal elements, are significant in magnitude. This latter case corresponds to an "unsharp" or "weak" measurement, in which the interaction between the quantum system and the measurement apparatus does not yield a good correlation for the system observable under study.

Let us now consider a typical example to illustrate the above: a two-slit experiment. The slits are labeled A and B, and the states of the quantum system corresponding to "passage through slit A" and "passage through slit B" are $|A\rangle$ and $|B\rangle$. These two orthogonal states form a basis for the two-dimensional system Hilbert space. The measurement apparatus pointer states corresponding to the above system state are $|x_A\rangle$ and $|x_B\rangle$. These form an orthonormal basis for the pointer Hilbert space. When the pointer is read, the outcome is either x_A or x_B, that is, an eigenvalue of one of the pointer eigenstates.

In the usual ideal or sharp measurement, we begin with the pointer in an initial ready state $|x_0\rangle$, and the evolution for each system state is as follows:

$$|A\rangle|x_0\rangle \rightarrow |A\rangle|x_A\rangle$$
$$|B\rangle|x_0\rangle \rightarrow |B\rangle|x_B\rangle. \tag{6.27}$$

If the system starts out in some arbitrary superposition of the basis states such as $(c_A|A\rangle + c_B|B\rangle)$, we find the evolution:

$$(c_A|A\rangle + c_B|B\rangle)|x_0\rangle \rightarrow c_A|A\rangle|x_A\rangle + c_B|B\rangle|x_B\rangle. \tag{6.28}$$

Now, suppose we instead choose a weak coupling between the system and the pointer, so that we have a "weak" or "unsharp" measurement. In this situation, the evolution for the system states $|A\rangle$ and $|B\rangle$ gains an error component such that the pointer has an amplitude, say, e, to give the wrong answer. For example, the pointer can yield the outcome $|x_B\rangle$ even though the system is in state $|A\rangle$. Of course, the pointer also has an amplitude for the correct result: call it d, where $|d^2| + |e^2| = 1$. The evolution for the system states under these conditions is

$$|A\rangle|\varphi_0\rangle \rightarrow |A\rangle|\varphi_A\rangle = |A\rangle(d|x_A\rangle + e|x_B\rangle)$$
$$|B\rangle|\varphi_0\rangle \rightarrow |B\rangle|\varphi_B\rangle = |B\rangle(e|x_A\rangle + d|x_B\rangle). \tag{6.29}$$

For an arbitrary superposition as in (6.28), the evolution is

$$(c_A|A\rangle + c_B|B\rangle)|\varphi_0\rangle \rightarrow c_A|A\rangle|\varphi_A\rangle + c_B|B\rangle|\varphi_B\rangle$$
$$= c_A d|A\rangle|x_A\rangle + c_A e|A\rangle|x_B\rangle + c_B e|B\rangle|x_A\rangle + c_B d|B\rangle|x_B\rangle. \tag{6.30}$$

From the total density matrix from the final state in (6.30), that is, $\rho = |\Psi\rangle\langle\Psi|$, we trace over the pointer to get the reduced density matrix for the quantum system. We find (with $i, j = \{A, B\}$):

$$\rho_S = \sum_{i,j} c_j^* c_i \langle\varphi_j|\varphi_i\rangle|i\rangle\langle j| = \begin{bmatrix} |c_A|^2 & c_A^* c_B \langle\varphi_A|\varphi_B\rangle \\ c_B^* c_A \langle\varphi_B|\varphi_A\rangle & |c_B|^2 \end{bmatrix}. \tag{6.31}$$

Note that $\langle\varphi_i|\varphi_j\rangle = \langle\varphi_j|\varphi_i\rangle = d^*e + e^*d$. If the error amplitude were to vanish, $e = 0$, we would have an ideal or sharp measurement of the "which slit" observable, and the system's reduced density matrix would be diagonal with respect to that basis. On the other hand, if $d = e$, that is, if the error is maximal, then the decoherence function is unity; that is, the relative pointer states coincide. (This follows from the normalization constraint $|d^2| + |e^2| = 1$.) In this case, we get no information at all from the measurement, since no matter what state the system is in, the pointer always yields the same result. It is completely insensitive to the system. Indeed, as we will see later, in the case of maximal error, no entanglement, occurs between the system and the measuring apparatus; there is zero correlation. Thus, a maximally weak measurement is not a measurement at all.

We've seen above that the inner product of the relative pointer states (decoherence function) specifies the resolution of the measurement interaction with respect to the system observable under study. An inner product of zero yielding a diagonal reduced density matrix indicates that the possibility of a wrong answer from the measurement pointer vanishes. That is, we always find a pointer outcome that corresponds reliably to an eigenvalue of the system observable. But as observed above, under the assumption of unitary-only evolution, the system's reduced density matrix is an improper mixture. This theoretical representation cannot be interpreted as reflecting a situation in which the system is now in a particular eigenstate of the measured observable (even if unknown). That is, it does not license an epistemic interpretation. Thus, under the unitary-only assumption, there is a discrepancy between the theoretical representation and the empirical finding of a definite eigenvalue.

We will now consider the second aspect of decoherence introduced above, which we termed (2): repetition of the measurement interaction. Consider a basic measurement interaction with significant error amplitude e, such that the system's reduced density matrix retains sizable off-diagonal elements. This indicates some "quantum coherence," that is, the ability the display interference effects with respect to a basis such as "slit A" and "slit B," so that fringes could be seen. A typical example is the emission of a photon from an atom, where the photon's wavelength is too long to provide very good localization of the atom. However, when the same interaction is repeated, we find that the off-diagonal elements decrease. The theoretical account of this process begins by assuming that each such measurement interaction is independent. In the conventional decoherence program, this assumption is more a convenience than an independently established fact or theoretical result. However, it turns out to gain justification in TI, as we will see in the next section.

We can obtain a differential equation for the rate of change of the off-diagonal elements ρ_{mn} due to the repeated measurement interaction. In order to do this, we make use of the rate of repetition Γ of the interaction (such as an emission rate), as well as the decoherence function $\langle \varphi_m | \varphi_n \rangle$. We define a parameter $\lambda = \Gamma(1 - \langle \varphi_m | \varphi_n \rangle)$ and note that the rate of change is given by

$$\frac{\partial \rho_{mn}}{\partial t} = -\lambda \rho_{mn}. \tag{6.32}$$

The solution to this equation gives us the off-diagonal elements as a function of time:

$$\rho_{mn}(t) = \rho_{mn}(0)e^{-\lambda t}. \tag{6.33}$$

Thus, the off-diagonal elements decrease exponentially with repeated measurement interactions, where the rate of decrease is governed by the parameter λ. However, under the usual unitary-only assumption, the improper mixture of the rapidly diagonalizing density matrix does not reflect determinacy of outcome, and the standard approach therefore is not consistent with empirical results. In the next section, we will see how the transactional picture remedies this shortcoming.

6.5.2 RTI Completes the Decoherence Account

As noted in previous chapters, the interaction between an emitter and a set of absorbers that gives rise to a real photon is just the Von Neumann "measurement transition," or "Process 1":

$$|\psi\rangle \rightarrow \sum_k |\langle k|\psi\rangle|^2 |k\rangle\langle k|. \qquad (6.34)$$

In (6.34), each weighted projection operator represents an *incipient transaction*. The weights are the Born probabilities for the transferred photon's possible values of k (here, we suppress polarization indices). One of these becomes the *actualized transaction*, in which a real (on-shell) photon of momentum k is transferred from the emitter to the *receiving absorber*.

The transformation (6.34) takes a pure state to a mixed state that is diagonal in a specific basis, with respect to which one can define a Boolean probability space. That is, the probabilities over the possible outcomes sum to unity and obey all the usual Kolmogorov probability rules. One of the outcomes definitely occurs, with the associated probability; the others definitely do not. This means that the mixed state (6.34) describes a proper mixture; the statistical description describes a situation of epistemic uncertainty. The "propriety" of the mixed state arises because of the non-unitary transition precipitated by absorber responses.

In this typical example, the measurement basis is directional momentum. This quantity serves as an effective "pointer" in that it localizes the emitted photon to a particular absorber and also yields directional information about whatever particle emitted it, typically an atom or molecule under study. Thus, the process yields exact (as opposed to approximate), physically well-grounded decoherence. It also yields a naturally preferred basis, namely, that of conserved currents such energy/momentum. Recall that there is no well-defined position observable for photons, since they are ultra-relativistic, and there is no well-defined relativistic position observable.[24] It is also well known that there is no time observable even at the nonrelativistic level. In quantum field theories, it is standard practice to "demote" position to a mere parameter, along with time, in view of this fact. It is conserved quantities that are actually transferred in

---

[24] Technical details can be found, for example, in lecture notes by A. Neumaier (2016), who shows that certain relativistic quanta, including photons, "cannot be given natural probabilities for being in any given bounded region of space."

transactions, and that explains why these quantities constitute natural preferred observables, while the spacetime parameters do not.

Position only superficially appears to be a preferred or "natural" measurement basis because reduction or "collapse" usually takes place with respect to directional momentum (as discussed above), which singles out one micro-absorber. That absorber becomes excited, leading to amplified phenomena localized around that particular site, and at the macroscopic level, this appears as a kind of position "pointer." But the measurement really occurred with respect to momentum. In contrast to the typical unitary-only account, the physics of the transactional (direct-action) process thus leads to physically grounded localization and the natural emergence of classical determinacy. There is no need to circularly assume a classical "logical basis" or separable systems at the outset. Quantum systems are continually being separated from their entanglements because of the non-unitarity inherent in ubiquitous radiative processes of the kind represented in (6.34).

Now, recall from Sections 5.1 and 6.1 that the transactional account applies only to the transfer of massless bosonic gauge fields, that is, photons. The fermionic matter fields and bound states, such as atoms and molecules, are sources of the gauge fields. The sources themselves are not transferred by way of offers and confirmations in transactions. Instead, they participate as emitters and absorbers, that is, as the end points of transactions involving photon transfer. Importantly, a field source does not have to be directly detected/transacted in order to undergo "collapse" or reduction to a particular state. This is because sources are affected by the transaction involving the photon(s) they emit or absorb. It is the bound electron in an atom or molecule that directly serves as a source or sink of photons, but the entire bound state is affected by the process. An atom or molecule that participates, by way of its bound electrons, as a source or sink (i.e., emitter or absorber, respectively) in a transaction involving the transfer of a photon will certainly be affected by the transfer: it will undergo transformation to a state corresponding to the actualized emission of absorption of the photon.[25] For example, an atom prepared in a superposition of which-way states, upon emission of a photon of momentum k in an actualized transaction, will transition to a lower internal energy state as well as to a center-of-mass momentum state corresponding to that photon's momentum value. It will thus become localized, to a greater or lesser degree, to one of the which-way states. We will see a specific example of this process in what follows.

Let us now return to the simple two-slit example, assuming that the particle in the two-slit experiment is an atom that could emit one or more photons. As

---

[25] This is one way in which entire atoms can be detected in experiments probing their behavior. See Wilzbach et al. (2006) for some examples of the detection of neutral atoms using interactions with the electromagnetic field. An electron can also absorb a photon in an ionizing interaction and be liberated from a bound state. It can subsequently emit a photon in the inverse process, radiative recombination, to become part of a new bound state. The latter type of process serves to detect free electrons.

discussed previously, these photon emissions would constitute measurement interactions, because the emitted photons are correlated, to a greater or lesser extent, with the atom's "which slit" state. Thus, the photons serve as "pointers." However, in contrast to the usual unitary-only approach, "measurement" in TI is not just a unitary interaction; it includes a crucial non-unitary process corresponding to von Neumann's "Process 1." This process involves objective reduction to a definite outcome, even if that outcome may be unknown.

So, consider an atom prepared in an arbitrary state as in (6.28). If the atom is allowed to emit a photon, the combined system is in a state like (6.30), where we rewrite the photon state in terms of the correlated k values, consistent with the fact that there is no real position observable for photons:

$$(c_A|A\rangle + c_B|B\rangle)|\varphi_0\rangle \rightarrow c_A|A\rangle|\varphi_A\rangle + c_B|B\rangle|\varphi_B\rangle$$
$$= c_A d|A\rangle|k_A\rangle + c_A e|A\rangle|k_B\rangle + c_B e|B\rangle|k_A\rangle + c_B d|B\rangle|k_B\rangle. \tag{6.35}$$

Now suppose that there are two photon detectors D_A and D_B, corresponding to the atom's "which slit" states. This defines a two-dimensional photon subspace: $\{k_A, k_B\}$. However, we must note that (6.35) represents the total system state in terms of relative states for the photon; this prioritizes the atom's detection basis over that of the photon. We need to analyze the effect of the photon detections on the atom, which means we need to prioritize the photon detection basis. This will define the corresponding "unsharp" relative states for the atom, which will enable us to see what happens to the atom when its emitted photon is detected in a particular momentum state.

Accordingly, let us rewrite the total system state in terms of the photon basis to define the associated relative states $|\alpha\rangle, |\beta\rangle$ for the atom:

$$|\Psi\rangle = (c_A d|A\rangle + c_B e|B\rangle)|k_A\rangle + (c_A e|A\rangle + c_B d|B\rangle)|k_B\rangle$$
$$\equiv a|\alpha\rangle|k_A\rangle + b|\beta\rangle|k_B\rangle. \tag{6.36}$$

In (6.36), the normalized relative atomic states $|\alpha\rangle, |\beta\rangle$ are

$$|\alpha\rangle = \frac{1}{a}(c_A d|A\rangle + c_B e|B\rangle)$$
$$|\beta\rangle = \frac{1}{b}(c_A e|A\rangle + c_B d|B\rangle) \tag{6.37}$$

and the amplitudes a, b are defined as

$$|a|^2 = |c_A d|^2 + |c_B e|^2$$
$$|b|^2 = |c_A e|^2 + |c_B d|^2. \tag{6.38}$$

Note that (6.38) defines the amplitudes a and b of the atomic state components – the relative atomic states – that give rise to the different possible photon emission states $\{k_A, k_B\}$. In this sense, the atom itself acts as a kind of "beam splitter" in that its indeterminate state splits the generated photon OW into different components whose amplitudes correspond to the amplitude of that relative atomic state. Specifically, in this case, the atomic component $a|\alpha\rangle$ gives rise to a photon OW component $a|k_A\rangle$, while the other atomic component $b|\beta\rangle$ gives rise to a photon OW component $b|k_B\rangle$. Thus, what reaches the photon detector D_A is the photon component $a|k_A\rangle$. That prompts a confirmation $a^*\langle k_A|$, so that the resulting incipient transaction is represented by the outer product $|a|^2|k_A\rangle\langle k_A|$. In the usual manner, the weight $|a|^2$ – the Born Rule – is the probability of actualization of that incipient transaction (and *mutatis mutandis* for the component $b|k_B\rangle$).

What effect does actualization of one of these photon transactions have on the atom? If the photon transaction for $|k_A\rangle\langle k_A|$ is actualized, the photon's quantum state undergoes reduction, so that it conveys a determinate value of momentum k_A to its final absorber. This leaves the atom definitively in the state $|\alpha\rangle$ (and similarly for D_B). Thus, actualization of the photon emitted by an atom in an indeterminate state results in reduction of the atom's state to the component corresponding to that photon transaction. But of course this reduced atomic state still need not be an eigenstate of any particular center-of-mass observable – the relative atomic states are still superpositions of the which-way states, and this is why some degree of coherence of the atom can be retained, even if it emits a photon.

However, at this point, the total system is in some definite reduced state, whether or not we know what it is. Therefore, we can describe the total system in terms of a density matrix ρ that represents an epistemic mixture – since a definite photon outcome and corresponding reduction of the atomic state has occurred. This density matrix is a weighted sum of the incipient transactions available to the combined system of atom plus photon, that is,

$$\rho = |a|^2|\alpha\rangle\langle\alpha|\otimes|k_A\rangle\langle k_A| + |b|^2|\beta\rangle\langle\beta|\otimes|k_B\rangle\langle k_B|. \tag{6.39}$$

The corresponding reduced density matrix for the atomic subspace is then

$$
\begin{aligned}
\rho_S &= |a|^2|\alpha\rangle\langle\alpha| + |b|^2|\beta\rangle\langle\beta| \\
&= (c_A d|A\rangle + c_B e|B\rangle)(c_A^* d^*\langle A| + c_B^* e^*\langle B|) \\
&\quad + (c_A e|A\rangle + c_B d|B\rangle)(c_A^* e^*\langle A| + c_B^* d^*\langle B|) \\
&= \begin{bmatrix} |c_A|^2(d^*d + e^*e) & c_A^* c_B(d^*e + e^*d) \\ c_B^* c_A(d^*e + e^*d) & |c_B|^2(d^*d + e^*e) \end{bmatrix} = \begin{bmatrix} |c_A|^2 & c_A^* c_B\langle\varphi_A, \varphi_B\rangle \\ c_B^* c_A\langle\varphi_B, \varphi_A\rangle & |c_B|^2 \end{bmatrix}.
\end{aligned}
$$

$$\tag{6.40}$$

This is of course the same expression as in the standard decoherence approach for the system's reduced density matrix. However, in the transactional account, tracing over the photon pointer subspace now corresponds to taking into account the detection of the photon and its real (even if unknown) physical effect on the atom, while not specifying any particular detection scheme for the atom (i.e., no final observable for the atom is defined). Thus, (6.40) is a statistical description of epistemic uncertainty about the atom after it emits a photon that is actually absorbed, in a non-unitary process, at a particular detector. That is, the atom *really* acquires the state $|\alpha\rangle$ or $|\beta\rangle$ corresponding to a photon detection at D_A or D_B, so (6.40) is an epistemic mixture of those states. Thus, unlike in the conventional unitary-only account, we are not ignoring putative ongoing entanglement that yields an improper mixture for the atom. Rather, we now have a physically justified proper mixture – one that can be interpreted epistemically.

As noted above, owing to the error component, in general the relative atomic states $|\alpha\rangle$ and $|\beta\rangle$ are superpositions of the which-slit states, so the atom's density matrix retains nonvanishing off-diagonal elements indicating some degree of retained coherence. This coherence will of course influence the probabilities for outcomes of a specified atomic observable. For the special case of maximal error, $d = e = 1/\sqrt{2}$, interference fringe visibility is maximized. This case involves vanishing coupling between the atom's which-slit states and the photon, and the relative atomic states coincide to the initial prepared state: $|\alpha\rangle = |\beta\rangle = c_A|A\rangle + c_B|B\rangle$. We can see from (6.36) that for maximal error, no entanglement of the quanta is created. That is, for $d = e = 1/\sqrt{2}$ the resulting total state becomes

$$|\Psi\rangle = \frac{1}{\sqrt{2}}(c_A|A\rangle + c_B|B\rangle)|k_A\rangle + \frac{1}{\sqrt{2}}(c_A|A\rangle + c_B|B\rangle)|k_B\rangle$$

$$= (c_A|A\rangle + c_B|B\rangle)\frac{|k_A\rangle + |k_B\rangle}{\sqrt{2}} \equiv (c_A|A\rangle + c_B|B\rangle)|k+\rangle. \qquad (6.41)$$

Thus, for maximal error, the total system remains a product state, and the atom simply emits a photon in a "both slits" state, $|k_+\rangle$, which is insensitive to the atom's prepared center-of-mass state. Under these circumstances, no entanglement is created through the photon emission. This is an important point with relevance to the concept of "weak measurements," often claimed to yield special information about quantum systems. In fact, we see above that whatever information is gained about a system through the creation of a measurement correlation – that is, an entanglement – is always paid for by the cost of projecting the system into some new relative state such as $|\alpha\rangle$ or $|\beta\rangle$, which nontrivially disturbs it. The only way we can completely eliminate any disturbance of the system is by maximizing the error component. But if we do that, we gain no information at all, since there is no entanglement created and no measurement correlation established. Therefore, it is never tenable to assert that weak measurements leave a system undisturbed, even in some limit of vanishing coupling. If the system is truly undisturbed, there is no measurement at all; zero information is gained, as reflected in (6.41), in which the outgoing photon tells us nothing at all about the amplitudes c_A and c_B. A *maximally weak measurement is a nonmeasurement*, at least with respect to a center-of-mass observable such as the "which slit" observable. (Of course, if the atom emits a photon, we know that its internal energy state changed.)

Now let us consider the opposite limit, that of zero error ($d = 1$, $e = 0$). In this case, fringe visibility vanishes, because the relative atomic states are then the orthogonal "which slit" states, that is, $|\alpha\rangle = |A\rangle$, $|\beta\rangle = |B\rangle$. However, there is an important special case that commonly causes great confusion concerning the implications of the loss of fringe visibility. This is the state that features in so-called quantum eraser (QE) experiments, in which $c_A = c_B = 1/\sqrt{2}$. For equal coefficients, the entangled state becomes ambiguous as to pointer basis, since the state takes the same form when written in terms of the "both slits" basis, that is,

$$|\Psi\rangle = \frac{1}{\sqrt{2}}(|A\rangle|k_A\rangle + |B\rangle|k_B\rangle) = \frac{1}{\sqrt{2}}(|+\rangle|k_+\rangle + |-\rangle|k_-\rangle) \qquad (6.42)$$

$$\text{where} \quad |\pm\rangle = \frac{|A\rangle \pm |B\rangle}{\sqrt{2}} \quad \text{and} \quad |k\pm\rangle = \frac{|k_A\rangle \pm |k_B\rangle}{\sqrt{2}}.$$

Thus, we get exactly the same probability distribution as the which-slit basis, lacking interference. This is because there are two interference patterns corresponding to the states $|\pm\rangle$: "fringe" and "antifringe," which are exactly out of phase. It is only when one looks at sub-ensembles obtained by coincidence counting that one can find interference effects. This basis ambiguity underlies claims about "erasure" that are actually untenable, as discussed in the previous chapter. The crucial (and very common) error is to assume that loss of interference indicates a preference for the "which slit" basis and attendant existence of "which slit" information, when that is unjustified. For further details, see Kastner (2019b).

Let us now take a look at some specific examples of "weak" or "unsharp measurements" (i.e., with a nonvanishing error component e) in which we can find some fringe visibility on a final screen or equivalent detection setup. Such experiments involve a pointer basis consisting of some discrete position parameter, such as a set of pixels along a screen or a movable detector that can be set at different discrete positions x.

For the case in which $c_A = c_B = 1/\sqrt{2}$, the atom's relative states are

$$|\alpha\rangle = d|A\rangle + e|B\rangle$$
$$|\beta\rangle = e|A\rangle + d|B\rangle. \tag{6.43}$$

And since $a = b = 1/\sqrt{2}$, the total state, from Equation (6.36), is

$$|\Psi\rangle = \frac{1}{\sqrt{2}}[|\alpha\rangle|k_A\rangle + |\beta\rangle|k_B\rangle]. \tag{6.44}$$

The atom's reduced density matrix ρ_s is now:

$$\rho_s = \frac{1}{2}(|\alpha\rangle\langle\alpha| + |\beta\rangle\langle\beta|) = \begin{bmatrix} \frac{1}{2} & \frac{1}{2}\langle\varphi_A, \varphi_B\rangle \\ \frac{1}{2}\langle\varphi_B|\varphi_A\rangle & \frac{1}{2} \end{bmatrix}. \tag{6.45}$$

Recall that the atom's reduced density matrix now arises from a set of incipient transactions for the photon; that is, it is a sum of the weighted projection operators for each of the relative atomic states corresponding to their respective photon detections. One of these is actualized, and the photon is really detected at D_A or D_B, which projects the atom into the corresponding relative state $|\alpha\rangle$, $|\beta\rangle$. Thus, the weights accompanying the associated projection operators – that is, the Born probabilities – are just measures of our ignorance about which state the atom actually acquires as a result of the photon detection. This means that the atom's reduced density matrix (6.45) is a physically justified epistemic mixture.

Now, let us incorporate the unitary evolution of the atom's "which slit" states $|A\rangle$, $|B\rangle$ to the final screen or movable detector. This leads to amplitudes $\langle x|A\rangle$, $\langle x|B\rangle$ for detections at the positions x. The relative states in the X basis are

$$|\alpha\rangle = \sum_x (d\langle x|A\rangle + e\langle x|B\rangle)|x\rangle$$
$$|\beta\rangle = \sum_x (e\langle x|A\rangle + d\langle x|B\rangle)|x\rangle. \tag{6.46}$$

Recall from our previous discussion that detection of the atom is actually indirect, by way of secondary photon transactions. These typically involve scattering of an auxiliary photon "probe" beam by the atom, for example, via fluorescence. In what follows, we'll describe detection of the atom in terms of OW and CW, but strictly speaking, these OW and CW describe secondary photon transactions, in which the imposed probing photon's state becomes correlated with that of the atom. So the atom is really an intermediary whose state is probed through its interactions with the auxiliary photon beam, and it is the latter that actually participates in the transactions as the OW and CW components described below.

For an emitted photon detection at D_A, the emitting atom is projected onto the state α, so a particular pixel x receives the component

$$(d\langle x|A\rangle + e\langle x|B\rangle)|x\rangle. \tag{6.47}$$

It therefore responds with the adjoint confirmation

$$(d^*\langle A|x\rangle + e^*\langle B|x\rangle)\langle x|. \tag{6.48}$$

The associated incipient transaction is described by the outer product of the OW and CW,

$$|(d\langle x|A\rangle + e\langle x|B\rangle)|^2|x\rangle\langle x|. \tag{6.49}$$

Since for a nonvanishing error e we have a squared sum of amplitudes, the probability distribution exhibits interference fringes.

We get another set of incipient transactions, for each value of x, corresponding to the atomic state β correlated with photon detection at D_B. The resulting total probability distribution $P(x)$ on the screen will be the sum of both sets:

$$P(x) = \frac{1}{2}\left\{\left|(d\langle x|A\rangle + e\langle x|B\rangle)\right|^2 + \left|(e\langle x|A\rangle + d\langle x|B\rangle)\right|^2\right\}. \tag{6.50}$$

In (6.42), the two terms correspond to the atomic relative states α and β. Note that for a sharp measurement ($d = 1$, $e = 0$), we end up with the usual sum of the "which slit" probability distributions:

$$P(x)_{sharp} = \frac{1}{2}\left\{|\langle x|A\rangle|^2 + |\langle x|B\rangle|^2\right\}. \tag{6.51}$$

On the other hand, for maximal error, $d = e = 1/\sqrt{2}$, we find maximal fringe visibility because (as noted previously) the relative states α and β coincide, and we get:

$$P(x)_{unsharp} = \frac{1}{2}\left\{|\langle x|A\rangle + \langle x|B\rangle|^2\right\}. \tag{6.52}$$

The probabilities are of course the same as in the conventional unitary-only approach to quantum theory. But under the above analysis, we now have a physical reason for the squaring procedure of the Born Rule, as well as for the legitimate interpretation of the system's reduced density matrix as an epistemic mixture. Thus, we have arrived at the same reduced density matrix of the system under study as in the usual approach, but it is a proper mixture, as is needed for a theoretically consistent description of measurement results.

Let us return now to the second aspect of decoherence, that is, the vanishing of the off-diagonals with repeated measurements. At this point, the matter becomes trivial. The same basic analysis as in the previous section applies, but the measurement interaction involves non-unitary reduction. Thus, with each measurement, our proper mixture becomes more and more diagonal in the measurement basis defined by absorber responses (and the applicable Hamiltonian). An additional dividend of the transactional picture is that we now have a clear theoretical reason for the previously postulated independence of the measurements: specifically, the non-unitary collapse is what makes these independent. Moreover, at the relativistic level of RTI, we find that measurement events correspond to decay rates, which are known to satisfy Poissonian statistics (cf. Kastner and Cramer, 2018). Thus, the transactional picture provides a rigorous account of the onset of Markov behavior.

Finally, it should be noted that the above analysis does not preclude the phenomenon of "recoherence," as discussed (for example) in Bouchard et al. (2015). These authors note that a suitable inverse unitary operation can reverse the entanglement of several degrees of freedom. Such an analysis applies also under the transactional picture, which takes into account any unitary evolution that applies to the system ahead of a final absorption opportunity. If an appropriate inverse unitary process is imposed on the entangled degrees of freedom ahead of their interaction with absorbers, then those absorbers will of course respond to whatever pure states existed prior to the entanglement. Thus, recoherence is always possible if interaction with absorbers is deferred until after the inverse unitary process.[26]

Another way to understand the issue is by way of Markovian versus non-Markovian processes. A non-Markovian process is characterized by the preservation of unitarity. According to the transactional picture, whenever an opportunity for absorption is present, there is a chance for loss of unitarity (i.e., measurement and reduction), and then we have a Markovian process describable by a master or Lindblad equation. At that point, recoherence is no longer possible. It is important to note that genuine Markovian processes can arise only in the presence of real physical non-unitarity. The usual unitary-only approach to quantum theory precludes a rigorous, noncircular account of the onset of Markov behavior (see Kastner, 2017b). Thus the transactional picture offers a rigorous account of the emergence of empirically observed Markov behavior, while also allowing for empirically corroborated recoherence under appropriate unitary conditions.

6.5.3 Time Evolution in RTI

The example of an atom becoming decohered by way of photon emission raises the issue of an important distinction between the time evolution of rest-mass systems

---

[26] The temporal language here should be taken in a "process" sense rather than in spatiotemporal terms. This just means that the inverse unitary process must take place between the emission and any absorption opportunity.

and that involving photons. The prevailing approach to time evolution in quantum theory is the Schrödinger picture, in which the quantum state is assumed to be time-dependent while the observables are static quantities. Quantum systems with nonvanishing rest mass can be consistently described in this way, but photons cannot. The non-relativistic Schrödinger equation explicitly involves mass:

$$i\hbar\frac{\partial}{\partial t}|\psi\rangle = \left(-\frac{\hbar^2}{2m}\nabla^2 + V\right)|\psi\rangle.$$

Thus, the unitary evolution implied by the time-dependent Schrödinger equation is applicable to systems with finite rest mass, but not to massless photons, which are extreme relativistic objects. This situation requires another careful distinction between emitting and absorbing systems, on the one hand, and photons, on the other, in connection with measurement in RTI. As emphasized in the example discussed in this section, according to RTI, any quantum measurement necessarily involves photon transfer, including measurement of quanta with nonvanishing rest mass. That is, quanta such as atoms and electrons are *indirectly* measured by way of photon interactions resulting in photon transactions. This is why observables are generally described by projection operators, that is, by a "spectral decomposition": such a description reflects a set of incipient transactions denoting the different possible photon outcomes corresponding to eigenvalues of the observable.[27]

As illustrated previously in this section, an atom can be prepared in a superposition of "which path" states, and detected in an eigenstate of a "which path" observable based on the outcome of a photon transaction involving the atom acting as an emitter or absorber. The observable, say, W, would be described by the spectral decomposition $W = a|A\rangle\langle A| + b|B\rangle\langle B|$, where a, b are eigenvalues for "detection in slit A, B." This is an observable for a measurement of an atom, but its form as a sum of projection operators comes from the necessary involvement of the electromagnetic field in providing the transactional process (offers and confirmations) leading to indirect measurement (reduction) of the atom's resulting state. That is, there is no "atom confirmation." Rather, the atom is described by a quantum state that undergoes the unitary Schrödinger evolution until it participates in a photon-mediated transaction, which then collapses its state in correspondence with the photon's absorption that triggers a particular outcome (as illustrated in the example discussed in previous subsections).

The reader might worry: Doesn't the time index applying to the atom's state single out an arbitrary preferred frame and thus lack relativistic covariance? This issue is discussed more fully in Chapter 8, but the short answer is that it does not because the atom's proper time is based essentially on its DeBroglie rest

[27] An exception is the number operator of quantum field theory.

frequency, which serves as the fundamental reference for any such temporal index applying to a quantum system with finite rest mass. Thus, the relevant temporal index is identified not through any preferred frame, but rather by reference to the proper time of a specific object. This can be thought of as an "internal time," in that it applies to a quantum object, not to any metrical property of spacetime.

On the other hand, when dealing with measurements of photons, the Heisenberg picture of measurement becomes fully applicable. Specifically, the photon state has no time evolution. The photon transfer establishes a null interval. There is no applicable proper time, and no corresponding inertial frame. In this case, any time evolution applies not to the photon's state but to the relevant observable. So for a measurement of some observable A on a photon, any time evolution would attach directly to A and would arise from $(\partial A/\partial t) = (i/\hbar)[H, A]$ where H is the applicable Heisenberg Hamiltonian.[28] In this case, the time index pertains to the inertial frame in which H is defined, and is still a fully covariant quantity that transforms according to the dictates of relativity theory. In Chapter 8, we will see in more detail how inertial frames arise in the RTI ontology.

6.6 RTI Solves the Frauchiger–Renner Paradox

The Frauchiger–Renner (FR) paradox (2018) is a recent thought experiment based on an extension of the Schrödinger's cat and Wigner's friend paradoxes.[29] This class of paradoxes involves macroscopic superpositions that are never empirically observed. Such paradoxes arise only under unitary-only quantum theory, that is, the conventional approach to the theory that assumes that all physical evolution involving quantum systems is unitary. In what follows, we will designate this conventional approach as "UO QM." However, for present purposes we also include so-called textbook quantum theory, which includes the "projection postulate" (PP) as a form of UO QM. Under von Neumann's classification (von Neumann, 1932), the unitary evolution was termed "Process 2" while the non-unitary PP was termed "Process 1." However, even though it is mathematically non-unitary, the PP of textbook quantum theory is only a formal, ad hoc postulate that is not accompanied by any quantitative physical account, unlike that of the unitary evolution. So even quantum theory that includes the PP is physically unitary-only, since there is no quantitative physical model for the non-unitary evolution in the standard theory. Indeed, the ambiguity regarding the point in the analysis in which the PP should be applied is part of the measurement problem and a key ingredient in the derivation of the Frauchiger and Renner paradox. This

---

[28] It's assumed here that A has no explicit time dependence.
[29] This section is based on Kastner (2020b).

ambiguity is often viewed as unavoidable and is termed a "shifty split" or the ostensibly movable "Heisenberg cut" between the quantum system(s) under study and an external macroscopic world that is never independently defined.

The present classification of textbook presentations of quantum theory as unitary-only is perhaps nonstandard, but it is justified since, as noted above, the PP is not a physical account but only an arbitrary, ad hoc device. But for clarity, we can distinguish formulations that include the PP (but lack any physical account of non-unitarity) from Everettian or "many worlds"–type theories that deny collapse: let us denote the former as UOPP. In recent years, in view of the UO decoherence program which has often erroneously been taken to explain classical determinacy,[30] "textbook" UOPP quantum theory has become somewhat less fashionable, with many researchers gravitating to an Everettian view, either explicitly or tacitly. The ambiguity surrounding "whether to postulate projection or not" under an assumed linear-only dynamics is also a key aspect of the Frauchiger and Renner paradox.

In contrast, a form of quantum theory with genuine physical non-unitarity, such as the approach described in this book, contains a specific quantitative account of the non-unitary process and an unambiguous criterion for the transition between the domain described by quantum formalism and that describable by classical physics, such that there is no "shifty split," and the theory does not predict macroscopic superpositions. That criterion distinguishing the quantum from classical domains was specified in the previous chapter in Section 5.3.3. Specifically, the demarcation for emergence of classical determinacy is defined by circumstances under which we are virtually assured to have a NU-interaction (a transaction) involving our system of interest within the relevant spacetime interval. As described in Section 5.3.3, such circumstances are quantitatively defined under RTI, even though they are irreducibly indeterministic and thus governed by a probabilistic description.

It should be noted at this point that by far the most well-known "explicit collapse theory" is the Ghirardi–Rimini–Weber (GRW) theory (Ghirardi et al., 1986), and most researchers who have written about the Frauchiger and Renner paradox note that you can largely avoid it by using such a theory. However, the GRW theory is a different theory from quantum mechanics, since it modifies the usual linear evolution, adding an ad hoc nonlinear component in order to force collapse. In the literature on the Frauchiger and Renner paradox, authors apply a

---

[30] The UO decoherence program has been criticized as circular by the present author and others. See, for example, Kastner (2014b) and references therein. In any case, as discussed in the previous section, the results of the standard decoherence program can be derived in TI in such a way as to yield a full account of the emergence of determinate results.

"broad GRW brush" to non-unitary approaches and routinely take any explicitly non-unitary approach as denying the idea that quantum theory universally applies, one of the assumptions of the Frauchiger and Renner argument. But the transactional picture cannot be correctly classified as a GRW-type approach. It is a form of real quantum theory. It denies that "real" quantum theory is unitary-only, and it yields predictions completely empirically equivalent to the unitary-only theory at the level of the Born probabilities. Thus, under TI one can "have one's cake and eat it too": one can assent to the idea that quantum theory universally applies, that is, that the Born probabilities are universally correct, but one can dissent from the assumption that "real" quantum theory leads to the "absurd superpositions" of the Schrödinger's cat, Wigner's friend, and Frauchiger–Renner paradox.

6.6.1 The Frauchiger and Renner Paradox

Let us now turn to some specifics of the Frauchiger and Renner paradox. It is an elaboration of the "Wigner's friend" scenario. In this thought experiment, Wigner (W) stands outside a lab containing his friend (F), who is measuring a quantum system. F sees a definite result, but W (according to the UO assumption) supposedly must describe F as being in a superposition of states corresponding to the system being measured. The Frauchiger and Renner paradox extends this basic situation by having two F-level observers and two W-level "super-observers." The quantum systems under study by the F-level observers are subject to an interaction that allows the derivation of an explicit contradiction: the F-level observers must disagree with the W-level observers regarding the probability of a particular outcome for an observable that could in principle be measured. We will not go into the specifics of this construction here; readers interested in those details may find the discussion by Bub (2017) illuminating, although this author arrives at a different interpretation of the implications of this thought experiment.

Frauchiger and Renner develop their argument by reference to a friend, F, who is measuring a two-level spin system S with measuring device D in a lab L. They present the two levels of description, F and W, as follows:

- F assigns to his system S, after measuring its spin along z, either

$$|\uparrow\rangle_S \text{ or } |\downarrow\rangle_S. \tag{6.53}$$

- The external observer W assigns to S, D, and F the "absurd" overall laboratory superposition (the authors themselves use the term "absurd"):

$$|\Psi\rangle_L = \frac{1}{\sqrt{2}}\left(|\uparrow\rangle_s \otimes \left|\text{``}z=+\frac{1}{2}\text{''}\right\rangle_D \otimes |\text{``}\psi_S=|\uparrow\rangle\text{''}\rangle_F + |\downarrow\rangle_s \otimes \left|\text{``}z=-\frac{1}{2}\text{''}\right\rangle_D \otimes |\text{``}\psi_S=|\downarrow\rangle\text{''}\rangle_F\right).$$

$$(6.54)$$

Under these assumptions, and with the help of carefully chosen interactions, the authors obtain a contradiction: F-level observers will predict a probability of zero for an outcome that has a nonvanishing probability to be found by "super-observers" W. For our purposes, we don't have to go into the specifics of what this observable is; it's an observable applying to a macroscopic entangled system such as that described by (6.54). We will see in what follows that this sort of inconsistency is really not that surprising, since there is an initial fundamental inconsistency in the way in which the UO account is applied: (6.53) and (6.54) are already inconsistent. Indeed, it is the UO account, together with the ambiguity about the application of the PP, that gives rise to the contradiction. The moral of the story is that the UO account is fatally flawed, and what we need to do is recognize that quantum theory has genuine, physical non-unitarity, as described by TI.

6.6.2 Origin and Resolution of the Frauchiger–Renner Inconsistency

Let us first review the usual account of measurement under the unitary-only (UO) assumption. Consider an observable R with eigenvalues $\{r\}$, and suppose the system of interest, S, is initially in some arbitrary superposition $|\psi\rangle$ of eigenstates $|r\rangle$. The usual UO story is that performing a measurement on S consists of nothing more than the introduction of a correlation between S and a measuring device D. If D starts out in a ready state ϕ_0, the evolution looks like

$$|\psi\rangle|\phi_0\rangle \rightarrow \sum_r |r\rangle|\phi_r\rangle.$$

$$(6.55)$$

Suppose we subsequently find that D gives the result $|\phi_r\rangle$. It's important to note that UO QM does not allow us to attribute an eigenstate $|r\rangle$ to our system. This is because according to UO QM, such a system continues on as a component subsystem of a composite entangled state; this is mandated by the evolution (6.55). Indeed, according to (6.55) there is no account of our having "found" any particular result at all (this is just the measurement problem). This same point applies to UOPP, since under the PP there is no physical account of any real collapse or reduction to an outcome eigenstate such as $|r\rangle$. It is just an ad hoc postulate based on the fact that we always find some such outcome. There is nothing in the standard approach that explains *why* we found an outcome, and the standard approach, which specifies by (6.55) that entanglement remains, dictates that neither S nor D are in pure states – they are in improper mixed states (of the kind discussed in the previous section on decoherence). So, according to the standard approach (even enhanced with the PP), no matter what we find in the laboratory, there is no *physical* justification for attributing a pure state such as $|r\rangle$ to our system S.

Given the UO assumption that after measurement the total system is described by the right-hand side of (6.55), we are faced with the same issue discussed in the previous section: our system S is in an improper mixed state that cannot be interpreted as representing a situation in which S possesses a well-defined (even if unknown) value for the measured observable. Let us consider a simple example to see what kind of trouble we can get into by trying to treat an improper mixture as if it's a proper mixture amenable to an epistemic interpretation, in which our system really is in some definite eigenstate of the measured observable.

Suppose we have two experimenters, Alice and Bob, working with two electrons in a typical EPR-Bell state, such as the "triplet" $s = 1$, $m = 0$, state:

$$|\Psi\rangle = \frac{1}{\sqrt{2}}(|\uparrow\rangle|\downarrow\rangle + |\downarrow\rangle|\uparrow\rangle). \tag{6.56}$$

Now, suppose that Alice measures electron A along the spin direction z, while Bob measures electron B along some other spin direction θ. Each will find the electron either "up" or "down" in the direction they measured. If each of them assigns to their measured electron the eigenstate corresponding to the result they found, the resulting possible total state assignments will be as follows:

$$
\begin{aligned}
&|\uparrow\rangle_z \otimes |\uparrow\rangle_\theta \\
&|\uparrow\rangle_z \otimes |\downarrow\rangle_\theta \\
&|\downarrow\rangle_z \otimes |\uparrow\rangle_\theta \\
&|\downarrow\rangle_z \otimes |\downarrow\rangle_\theta.
\end{aligned}
\tag{6.57}
$$

This would mean that, from the vantage point of an outside experimenter, say, Walter, who did not know what those outcomes were, the composite system would have to be in the *proper* mixed state:

$$\rho = \sum_{i,j} \Pr(i,j)|i\rangle_z \langle i|_z \otimes |j\rangle_\theta \langle j|_\theta \tag{6.58}$$

where $i, j \in \{\uparrow\downarrow\}$, and $\Pr(i, j)$ is the probability for each of the states in (6.57). *But this mixed state contradicts the state (6.56), which is a pure state.* A pure state and a proper mixed state are completely different animals, leading to completely different statistical predictions for certain observables.

Now, this inconsistency is generally treated as unproblematic, based on the idea that (under UO) the electrons will become entangled with environmental degrees of freedom. This putative ongoing entanglement will suppress interference pertaining to observables corresponding only to the electrons, such as the angular momentum of the triplet state. Thus, so the argument goes, Walter would find the kinds of statistics corresponding to the mixed state (6.58). However, this is only a "FAPP" account, since the UO theory still rules out the state (6.58). But even if we accepted this "entanglement with additional degrees of freedom" loophole as an explanation for why Walter would see the statistics corresponding to (6.58), it would not be of any use for resolving the Frauchiger and Renner inconsistency. This is because the "super-observable" measured by the "super-observers" at the W level involves *all* the relevant degrees of freedom (up to and including *F* himself). So the usual loophole appealing to ongoing entanglement with additional degrees of freedom is no longer available.

The upshot of the above is that under the UO assumption, the outcome-based states (6.53), analogous to (6.58), cannot physically describe the system nor can they yield reliably correct predictions for putative macroscopic superpositions such as (6.54). But the Frauchiger and Renner paradox depends on allowing some observers, like *F* above in state assignment (6.53), to assign to their measured system eigenstates corresponding to the outcomes they observed, yielding a "collapsed" state description analogous to (6.58). Since this is inconsistent with the UO assumption, it naturally leads to the inconsistency that manifests as disagreement about the probability of an outcome at the level of the "super-observers."

Frauchiger and Renner describe all their experimenters as applying quantum theory "correctly," but under UO, *F* is in fact *not* using quantum theory correctly. *F*'s state assignments (6.53) contradict the composite entangled state (6.54), just as (6.58) contradicts (6.56). In this regard, Frauchiger and Renner actually succumb to a fallacy. They say:

Although the state assignment [6.54] may appear to be "absurd," it does not logically contradict [6.53]. Indeed, the marginal on S is just a fully mixed state. While this is different from [6.53], the difference can be explained by the agents' distinct level of knowledge: F has observed z and hence knows the spin direction, whereas W is ignorant about it.

But as we saw above, the ignorance-based (epistemic) interpretation of improper mixed states is inconsistent and untenable. The Equation (6.54) *does* logically contradict (6.53), and the difference cannot be explained away in terms of ignorance, since improper mixed states are not legitimately amenable to an epistemic interpretation. If we ignore this contradiction and apply the collapsed state description at the *F*-level anyway, while also holding onto the uncollapsed description at the "super-observer" or *W* level, this initial inconsistency is what naturally manifests in the derived inconsistency down the line. That output inconsistency is the Frauchiger and Renner "paradox," but it is not a real paradox. It's just a result of letting one set of observers use a state attribution that is forbidden according the "rules of the UO game."

So how do we resolve the paradox? As indicated earlier, we reject the UO assumption, noting that under the transactional picture, real quantum theory in fact has a non-unitary component. This is borne out by the fact that in practice, physicists routinely assign outcome-related eigenstates to their measured systems, and never find these sorts of inconsistencies. This is not because they are applying a "projection postulate" without any real physics behind it. Rather, it's because in the real world, under real quantum theory – that is, in the transactional picture – we never really arrive at "absurd superpositions" like (6.54), because non-unitary reduction occurs well before that point. As discussed in Section 5.3.3, RTI naturally yields a stable and well-defined demarcation between the quantum level, in which systems are not amenable to a "collapsed" description corresponding to measurement outcomes, and the classical level, in which they are.

6.7 The Afshar Experiment

The Afshar experiment (2005) is an interesting variant on the famous two-slit experiment. In this experiment, a grid is placed at the location of the dark fringes in what would be an interference pattern, but, downstream from that location, the photon beam is subjected to a "which slit" measurement by way of lenses. However, its alleged significance is based on ambiguous and misleading terminology, and it has been widely misunderstood (and its significance has been consequently widely overstated). Afshar and his collaborators claim that the experiment challenges fundamental principles of quantum theory, such as the idea that one cannot obtain outcomes for two incompatible observables in a single measurement. The latter idea was pointed to by Niels Bohr in his "principle of complementarity" (POC), which Afshar claims to have refuted by his experiment. This section attempts to update and to correct the record, since key claims regarding the experiment's significance have in fact long ago been refuted. It is also a nice example for the application of TI.

6.7.1 Essence of the Afshar Experiment

I have been a critic of many of Bohr's pronouncements about quantum theory and have argued that his POC fails to pass muster as an interpretive principle (Kastner, 2016b). However, I did analyze the Afshar experiment in Kastner (2005), showing that there is in fact nothing at all paradoxical about it and that it does not refute the basic quantum principle that one can't obtain values for incompatible observables in a single measurement. Part of the ambiguity in assessing what bearing the experiment has on the POC is that Bohr's statement of the POC was somewhat sloppy. He noted that the experimental context determined what sorts of properties could be considered determinate, but didn't take into account than a single experiment can provide more than one such context. Thus, in a strict reading of the POC, such as "one cannot obtain results belonging to incompatible observables in a single experiment," one could retain the idea that the Afshar experiment presents a violation. But for consistency, one also would then have to assess a perfectly commonplace spin experiment involving a sequence of measurements of noncommuting observables as just as much a refutation, which would be absurd. This section presents a review of the experiment and shows that there is no paradox and no fundamental violation of the essence of the POC, which is just the idea that a quantum system does not exhibit properties of incompatible observables under a single measurement.

The Afshar experiment, schematically pictured in Figure 6.9, prepares a photon beam in a superposition of slit states corresponding to the upper and lower slits, that is,

$$|\psi\rangle = \frac{1}{\sqrt{2}}(|U\rangle + |L\rangle). \tag{6.59}$$

Then, at location σ_1, it interposes a grid placed exactly at the *minima* (i.e., areas of zero probability) in the predicted interference pattern for that prepared state based on the unitary evolution of the states to positions x. Just after the grid is a lens that focuses the components corresponding to each slit so that the "which slit" states are separated, and a final detection screen at σ_2 detects the probability distribution corresponding to those states – that is, separated spots rather than an interference pattern. The fact that the intensity at the final screen is not diminished shows that no photons were intercepted by the grid, and thus indirectly shows that there was an interference-type distribution at σ_1. Afshar claims therefore that we have both wavelike interference and particle-like "which way" phenomena in the same experiment, and that this violates Bohr's POC.

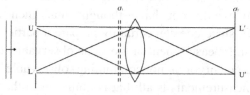

Figure 6.9 Schematic diagram of the Afshar experiment.

However, all that happens in the experiment is as follows: we subject the photon beam to a nondisturbing "null" measurement that confirms the prepared state (this is the grid placed at the dark areas for the predicted interference pattern), and then we subsequently subject the same photon beam to a "which slit" measurement. This process is fully equivalent to first preparing a spin-1/2 particle (such as a silver atom) in a state of "spin up along x," next confirming that it is in that state by having it pass through an x-oriented Stern–Gerlach device with a detector only for the state "spin down along x" (which never activates, since the particle was prepared in "spin up"), and finally allowing the particle to continue on to a z-oriented Stern–Gerlach device which detects it in either "spin up along z" or "spin down along z."

To see this explicitly, note that the state (6.59) is just a particular superposition of basis states in a two-dimensional Hilbert space. The Hilbert space for a spin-1/2 particle is also two-dimensional. For any given spin direction, we can represent the states "up" and "down" in terms of a computational basis commonly taken as the Z-direction. Thus, the states "up and down along Z" are analogous to $|U\rangle$, $|L\rangle$, respectively. With Z as the computational basis, the state "up along X" is completely analogous to the both-slits state (6.58). The grid performs the analogous function of a Stern–Gerlach device with a detector only for the state orthogonal to (6.59), that is, for

$$|\psi\rangle' = \frac{1}{\sqrt{2}}(|U\rangle - |L\rangle), \qquad (6.60)$$

so this detector will never be activated, and the prepared state will continue on unaffected. For the Afshar setup, the state (6.60) yields an "antifringe" interference pattern exactly out of phase with that of the prepared state (6.59). That is, the dark areas in the fringe pattern are bright areas in the antifringe pattern, and vice versa. The grid is placed at the dark areas in the fringe pattern, so it can only detect photons prepared in the orthogonal state (6.60) corresponding to the antifringe pattern. Thus, the grid is never activated as a detector. Returning to the spin-1/2 situation, it is then a simple matter to apply a Z-oriented Stern–Gerlach device to subject the prepared state (6.59) to a final measurement of Z, with the result that half the particles will be detected in the state "up along Z" (analogous to $|U\rangle$) and half will be detected in the state "down along Z" (analogous to $|L\rangle$).

Presumably, Bohr would not have thought that such a spin experiment contradicted his POC, since one can certainly prepare a particle in a state, confirm that state through a null measurement of the same observable, and then subject the particle to a measurement of a noncommuting observable. That sequence of perfectly mundane measurements is all that is going on in the Afshar experiment, as elegant an experimental exercise in quantum optics as it may be. The only criticism of Bohr's POC that one might make here is that he didn't state it very precisely; he sometimes stated it in terms of an "experimental setup" or a "single experiment." He really intended the idea that one cannot obtain values for incompatible observables in a *single measurement*, not in a single *experiment*, because of course one can do a sequence of noncommuting measurements in a single experiment.

6.7.2 *The Afshar Experiment in TI*

In the transactional picture, a photon is only created because emitters and absorbers interact in the special way that we termed a "NU-interaction" in the previous chapter. This usually involves a single emitter and several (or many) responding absorbers. Each responding absorber sets up an incipient transaction with the emitter. For the case of a single photon, only one of those responding absorbers actually receives the transferred photon conveying real conserved quantities such as energy, momentum, and angular momentum.

In the Afshar experiment, a set of incipient transactions arises between the final screen (at location σ_2 in Figure 6.9) and the photon source, where each incipient transaction corresponds to a micro-absorber in the screen (such as an atom). The nature of each incipient transaction depends on the offer wave (OW) component reaching each absorber in the screen, and the corresponding confirming (CW) responses. In the Afshar experiment, the OW that passes the slits is prepared in the "both slits–fringe" state. Unitary interaction of the OW with the lens just after σ_1 introduces a correlation between directional momentum and the "which slit" state. In conventional terms, this correlation makes the "which slit" components distinguishable after the lens (but not before). In TI terms, the lens acts on the photon momentum such that the OW components reaching absorbers in region U' in the final screen are much more likely to correspond to the directional momentum component corresponding to the U slit than to the L slit, and vice versa. Thus, these incipient transactions are all "which slit"–type transactions.

However, as noted above in connection with the conventional account, the slit-basis states are *not* distinguishable in the region between the emitter and the lens. In terms of TI, the OW between the emitter and the lens is still a "both slits" OW. With the grid placed at σ_1, there is no effect on the OW since every absorber in the grid corresponds to a region of zero amplitude – that is, no absorber in the grid receives any component of that OW. Thus the grid does not respond, and no incipient transactions can be set up between the emitter and the grid. This means that all incipient transactions between the emitter and the final screen are "which slit"–type transactions. Yet the presence of the grid, and the fact that no transactions are set up by the grid (i.e., no photons are intercepted at the grid), shows that the OW is indeed a "both slits" OW. But this does not mean that the final measurement is not a good "which slit" measurement, any more than a final measurement of "spin along z" with a result of "up along z" is not a good one for a particle prepared in "spin up along x."

In the next chapter, we consider in more detail the interpretation of quantum possibility, and how this approach can serve to resolve some long-standing tensions between realist and antirealist views about science and our understanding of the world.

7

The Metaphysics of Possibility in RTI

> All the world's a stage,
> And all the men and women merely players:
> They have their exits and their entrances;
> And one man in his time plays many parts ...
>
> —Shakespeare, *As You Like It*

The relativistic transactional interpretation in its possibilist form is a realist interpretation that takes the physical referent for quantum states[1] to be ontologically real possibilities existing in a pre-spacetime realm, where the latter is described by Hilbert space (or – more accurately – Fock space, accommodating the relativistic domain). We can characterize these possibilities as tokens of a new metaphysical category – *res potentia*. They are taken as real because they are physically efficacious, leading indeterministically to transactions which give rise to the empirical events of the spacetime theater.

For the remainder of this chapter, we'll again refer to the model as "PTI" to emphasize this particular ontology, but the reader should keep in mind that both names – RTI and PTI – refer to the same model in terms of its quantitative and formal features. The model can be considered in a weaker, agnostic, "structural realist" version, in which the Hilbert space structure of the theory is taken as referring to some structure in the real world without specifying what that structure is. In this form, one can retain the relativistic formal developments of RTI without endorsing the specific ontological interpretation of the referents of quantum states as *res potentiae*. (I specifically address the structural realism aspect in Section 7.6.) PTI in its strong form is very different from the traditional "possibilist realism" or "modal realism" pioneered by David Lewis. In order to make this distinction clear, I first briefly review the traditional account.

---

[1] The term "semantic realism" is often used to denote the idea that theoretical terms refer to specific physical entities, the position I advocate herein concerning quantum theory. In contrast, "epistemic realism" denotes the idea that we have good reason to believe a theory's claims. I consider a stance of epistemic realism about quantum theory as relatively uncontroversial, so I do not address it here.

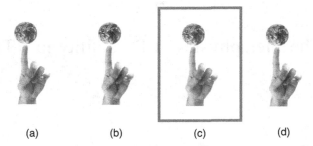

| (a) | (b) | (c) | (d) |

Figure 7.1 A set of "possible worlds" in traditional Lewisian possibilist realism. Worlds (a), (b), and (d) are possible worlds; the "actual world" (c) (in rectangle) is defined only relative to an observer. Each world is considered to be a complete, universal set of events.

7.1 Traditional Formulations of the Notion of Possibility

As noted in Chapter 1, David Lewis pioneered realism about possibilities in a comprehensive and sustained philosophical examination of entities he termed "possible worlds" (Lewis, 1986). In Lewis' formulation, possible worlds are the same sorts of entities as our own world. They are states of affairs that could conceivably occur, but which differ from the set of events in the actual (experienced) world. According to Lewis, these worlds are every bit as real as the actual world; the only difference is that the actual world is the one we happen to inhabit. Thus, in this theory, "actual" is *indexical*, meaning that it is a matter of perspective, not of kind or nature. Figure 7.1 illustrates this relationship schematically between Lewisian possible worlds and the actual world.

The Lewisian formulation is readily applicable to "many worlds"–type interpretations, in which each measurement event[2] causes a "branching" or copying of a particular world or collection of objects. However, PTI's proposed dynamic possibilities are fundamentally different from those of the Lewisian picture, as will be discussed in the next section.

7.2 The PTI Formulation: Possibility as Physically Real Potentiality

As noted above, Lewisian possible worlds are just alternative universal states of affairs and are no different in their basic nature from the actual world. In contrast, the dynamical possibilities referred to by state vectors in PTI are Heisenbergian *potentiae*, which are less concrete than events in the actual world, yet more

---

[2] Recall that the notion of a "measurement event" is ill-defined in Everettian interpretations because it requires dividing the physical objects under study into those which constitute the "measured system" and the "measuring apparatus." Such a specification is non-unique and therefore requires reference to an external observer or arbitrary choice.

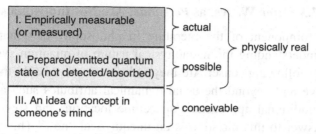

Figure 7.2 Quantum entities are less concrete than empirically measurable events, but more concrete than thoughts or merely conceivable situations.

concrete than mere thoughts or imaginings or conceivable events. This relationship is illustrated in Figure 7.2.[3] In contrast, as noted in Chapter 1, traditional approaches to measurement in quantum theory inevitably end up needing to invoke an "observing consciousness" in order to "collapse" the wave function (or state vector) and bring about a determinate outcome, necessitating speculative forays into psycho-physical parallelism. Thus, PTI is actually *less radical* than these much more common approaches because it does not need to invoke mental substance in order to address what certainly started out as a purely physical, scientific question about material objects.

Under PTI, the realist use of the term "possible" or "potential" refers to physical possibilities, that is, entities which can directly give rise to specific observable physical phenomena based on an actualized transaction.[4] This is distinct from the common usage of the term "possible" or "possibility" to denote a situation or state of affairs which is merely conceivable or consistent with physical law. So, in general, "possibilities" in PTI are entities underlying specific *individual* events rather than collective, universal sets of events such as the worlds in Figure 7.1. As an example, the possibilities underlying, for example, the detection of a photon at point X on a photographic plate are the offer wave components constituting the path integral in Feynman's "sum over paths" (recall Chapter 4).

Specific examples of each metaphysical category illustrated in Figure 7.2 are:

I. A detector click.
II. A spin-1/2 atom prepared in a state of "up along x."
III. "That possible fat man in the doorway."[5]

---

[3] Actually, mental activity *could* be considered real as well in that it could be based on quantum possibilities; this remains an interesting metaphysical question, but it is not crucial for PTI.

[4] This is very similar to, indeed perhaps the same as, Teller's proposal (Teller, 1997, 2002) that (however negatively stated in the words of Frigg, 2005, p. 512), the quantum field "has only something like structural efficacy, meaning that it does no more than [specify] the structure of physically possible occurrences."

[5] This is a reference to a famous 1948 paper by Quine, "On what there is," in which he criticizes traditional possibilist realism because of the apparent proliferation of any conceivable entity in a "slum of possibles" that is a "breeding ground for disorderly elements" (reprinted in Quine, 1953).

7.3 Offer Waves, as *Potentiae*, Are Not Individuals

A significant component of the literature in philosophy of quantum theory is addressed to understanding the metaphysical nature of quantum systems such as electrons in the following sense: Are they individuals, that is, do they have some "essence" above and beyond the usual dynamical attributes such as momentum, spin, and (in traditional approaches) spacetime location, and so on? In the PTI picture, the answer to this question is an unequivocal "no."[6] This is because the PTI (as well as original TI) ontology has no "particles" to whom one could even begin to attribute individualized "essences" or identities. In Section 7.3.1 we will see that a direct consequence of the nonexistence of particles is that quantum states are restricted in their mathematical form to be either symmetric (meaning unchanged under an exchange of subsystem labels) or antisymmetric (meaning changing only by a sign under an exchange of subsystem labels), and must therefore be either bosons or fermions. This latter feature of the quantum mechanics of multiparticle systems is sometimes viewed as a curious fact in need of explanation.[7]

7.3.1 *Wave Function Symmetry Related to Nonexistence of Particles*

First, recall that standard quantum mechanics assigns to a quantum sourced at a specific location in the laboratory, at some time $t = 0$, a Gaussian wave function[8] depending on the amount of time elapsed since its emission. Such a wave function is illustrated schematically in Figure 7.3(a). Now, suppose two quanta of the same type are created at $t = 0$ (say, both electrons). If sufficient time has elapsed, the wave function for the two quanta looks like Figure 7.3(b); that is, there is significant overlap (cross-hatched region). The usual way of discussing this is to say that there is no way to know which particle is described by which wave function, and therefore one has to assume that the particles are indistinguishable, where their indistinguishability is contingent on the fact that wave functions can overlap. However, in the RTI/ PTI ontology, there are no "particles" associated with either wave function, independently of whether or not the wave functions overlap. This leads to a different, but arguably stronger, demonstration of the fact that quantum states must be either symmetric or antisymmetric, as we will see in what follows.

---

[6] Thus I agree with Teller's view (1997) that quanta lack "primitive thisness."
[7] In particular, O. W. Greenberg has explored the idea of "parastatistics" in which the quanta are neither bosons nor fermions.
[8] Again, offer waves are not restricted to being wave functions, which are committed to a particular basis (usually the position basis); but this is probably the most familiar and intuitively easy way to conceptualize the issue under study.

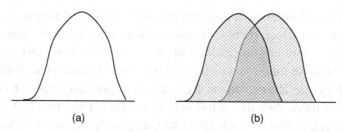

Figure 7.3 (a) The Gaussian wave function of a free quantum. (b) Overlapping wave functions of two free quanta.

The usual way of arguing that quantum states must be either symmetric or antisymmetric is by demanding that observable quantities (such as probabilities of detection) be invariant under a change of particle labels. For example, consider (as in Eisberg and Resnick, 1974) two particles in a one-dimensional box of side length a, one of them occupying the ground state $G(x_1)$ and the other occupying the first excited state $F(x_2)$, where x_1 and x_2 denote the location of each of the particles. (The two functions G and F have very different dependences on spatial location x.[9]) Now consider a non-symmetrized two-particle wave function such as $\Psi(x_1, x_2) = G(x_1)F(x_2)$. The probability density will be

$$P(x_1, x_2) = \Psi^*(x_1, x_2)\Psi(x_1, x_2) = G^*(x_1)F^*(x_2)G(x_1)F(x_2). \qquad (7.1)$$

But if we transpose the particle labels, then we get

$$P(x_2, x_1) = \Psi^*(x_2, x_1)\Psi(x_2, x_1) = G^*(x_2)F^*(x_1)G(x_2)F(x_1). \qquad (7.2)$$

In Equations (7.1) and (7.2) we have the functions G and F and their complex conjugates evaluated at different points x_i, so the probability densities $P(x_1, x_2)$ and $P(x_2, x_1)$ are not necessarily equal. In order to make them equal, we have to construct either the symmetric wave function Ψ_s or the antisymmetric wave function Ψ_A,

$$\Psi_S(x_1, x_2) = \frac{1}{\sqrt{2}}[G(x_1)F(x_2) + G(x_2)F(x_1)]$$

$$\Psi_A(x_1, x_2) = \frac{1}{\sqrt{2}}[G(x_1)F(x_2) - G(x_2)F(x_1)].$$

Thus, to review, the usual argument demands that empirically observable quantities such as the probability density be invariant under a transposing of particle labels based on the premise that quantum objects are "indistinguishable." The latter

---

[9] For this example, they are $G(x) \sim \cos \pi x/a$ and $F(x) \sim \sin 2\pi x/a$.

premise is arrived at because of an argument such as "wave function overlap makes it impossible to tell which particle is associated with which wave function."

Now suppose there are no particles at all. Then there is no auxiliary entity to associate with a wave function which could be "labeled," and which therefore could be addressed by the above sort of argument. But we can arrive at the need for symmetrization more directly as follows. Consider Equations (7.1) and (7.2). If there are no particles whose labels could be transposed, the only way to make these two expressions equal is to demand that $x_1 = x_2$. But if we do that, the resulting wave function refers to only *one* quantum. In the absence of auxiliary (labelable) quantum entities, the only way we can enforce the fact that there are two quanta is to provide two distinct arguments x_1 and x_2. Then the arguments *don't label anything*, but they are required in order to distinguish between a wave function for only one quantum and a wave function for two quanta. If they don't label anything, then there can be no physically appropriate meaning in an expression like $G(x_1)F(x_2)$, which implies a physical distinction between the two arguments of the functions G and F. The mathematical expression of the fact that *there is no physical distinction between the two arguments* is precisely the set of symmetric and antisymmetric wave functions above. Thus, *the observed fact that nature has only bosons (represented by symmetric states) and fermions (represented by antisymmetric states) can be arrived at simply by assuming that there are actually no "particles" (or individuals) meriting labels of any kind.* Again, we return to the idea that the fundamental ontological reality is that of nonlocalized fields and their excitations. The new feature proposed in PTI is that these fields represent possibilities for transactions, the latter corresponding to specific observable events.

7.4 The Macroscopic World in PTI

In this section, I consider macroscopic objects and the everyday level of experience in the transactional picture.

7.4.1 Macroscopic Objects Are Based on Networks of Transactions

I said in the previous section that there are no individualized "particles," just field excitations – Heisenbergian *potentiae* – that can lead to observable events via actualized transactions. Here I wish to address the question: What is it about transactions that make events "observable"?

First, recall that it is only through actualized transactions that conserved physical quantities (such as energy, momentum, and angular momentum) can be transferred. Such transfers occur between emitters and absorbers, which are fermionic field currents (recall Chapter 5). Thus the supporting entities and

Figure 7.4 Zooming in on a baseball.

structures for actualized transactions, as rest-mass systems, are "only" *potentiae* themselves. The realizing of phenomena is a kind of "bootstrapping" process in which actualized events are rooted in unactualized possibilities.

For a specific illustration, consider a baseball, depicted in Figure 7.4, as we zoom in to view it on smaller and smaller scales. The third square represents molecular constituents; the fourth square, a Feynman diagram, represents interactions among subatomic constituents both within molecules (intramolecular forces) and between molecules (intermolecular forces). A bound system such as an atom is a nexus of interacting forms of *potentiae* (field excitations such as those comprising nuclear quanta and electrons). But atoms and molecules can (and do) continually emit and absorb photons and other subatomic quanta. Those emitted quanta are absorbed by, for example, our sense organs, setting up enormous numbers of transactions transferring energy between ourselves and the atomic constituents of the baseball. The energy transfers effect changes in our brain, providing for our perception of the baseball.

Thus, in the transactional picture, a necessary feature and key component of any observation of a system is absorption of offer waves and corresponding generation of confirmation waves. We can go further and make a general interpretational identification of absorption with observation in a way not available to traditional interpretations of quantum theory: absorption is the way the universe "observes itself" and makes things happen. This identification is possible because absorption plays an equal role with emission in the dynamics of an event. In contrast, traditional interpretations take emission as the entire dynamical story and then cannot account for why observations seem to have such a special role in the theory. As Feynman tells us, we should sum the amplitudes over "unobserved" intermediate stages of an event to get a total amplitude for a final "observed" event, and then take the square of that. Why should we square that amplitude, and why should nature care whether we "observe" or not in this algorithm? *The only way that nature could know or care would be because something physical really happens in such "observations," and the only possible physical process accompanying an "observation" is absorption.* Under traditional interpretations that neglect absorption, the above apparently inexplicable procedure leads us into an impenetrable thicket of anthropomorphic considerations of the supposed effect of a mental substance – "consciousness" – on a physical substance, namely, a quantum system. In Feynman's words: "Do not keep saying to yourself, if you can possibly avoid it, 'But how can it be like that?' because you will get 'down the drain,' into a blind alley from which nobody has escaped. Nobody knows how it can be like that."[10] I suggest that an escape route from the "blind alley" is available; the price (or dividend, depending on one's point of view) is taking absorption into account as a real dynamical process and embracing the implications for our world view, which are explored in this and the next chapter.

7.4.2 Macroscopic Observation as Primarily Intersubjective

Next, let's consider a prototypical observation: once again, the two-slit experiment. Let's assume that the quanta under study are monochromatic (single-frequency) photons originating from a laser. In setting up the laser and the two screens, we handle macroscopic materials such as photographic plates. All of these actions consist of molecular-level transactions between enormous numbers of atoms and between some of the surface atoms and our hands. Energy is transferred via these transactions from those emitters to absorbers on our bodies; that energy serves as input for additional emissions between our sense organs and absorbers in our nerves, and so on, culminating in transfers of energy to our brains.[11] Brain changes make possible our perception that "something happened" (recall, from Chapter 2, Descartes' argument that it is not possible to observe anything that does not produce a perceptible change). But exactly *what happened* can vary considerably, depending on the specific transactions being actualized. A transaction between the photographic plate and my retina will not be the same as the transaction between another part of the plate and someone else's retina, but the laws of physics[12] ensure that all those many transactions are coordinated such that a coherent set of phenomena are created.

The point is that a macroscopic "observed event" is generally the product of an enormous number of transactions, even for only one observer. If one wishes to have one's observation corroborated, more transactions are required as another set of eyes, hands, and so on are introduced. These comprise a different set of absorbers, and the emitters may well be different as well. The transactions occurring for the second observer are not the same as those occurring for the first observer. For there to be corroboration, the two observers have to agree on

---

[10] Feynman et al. (1964).

[11] This description is not meant to be physiologically rigorous; it is merely an indication of how energy transfers via transactions ultimately result in brain changes.

[12] For example, conservation of physical quantities corresponding to the symmetries of the system and compliance with such laws as the principle of least action.

macroscopic facts such as "There is a dark spot at position $x = 50$," which can be instantiated by a large number of different sets of microscopic transactions. The process of corroboration is thus one of comparing the transaction-based perceptions of two (or more) different observers and deciding whether they represent the same macroscopic event. *But the macroscopic event itself can be no more than the sets of transactions taken as constituting it.* It is always definable only in terms of the subjective or intersubjective experiences of an observer or observers.

The above should not be taken as a reversion to mere subjectivism,[13] since for any individual transaction between emitter and absorber, there *is* an objective matter of fact concerning which transaction was actualized. Furthermore, there are certainly experiments in which an individual actualized transaction can be amplified to the macroscopic level, as in detection by a photomultiplier. But even in the case of amplification of a single transaction to the observable level, the type of event observed depends on what absorbers are present for the emitted quantum. In general, ordinary macroscopic events are collections of enormous numbers of transactions, with different sets of transactions for different observers.

7.4.3 Implications for the Realism/Antirealism Debate

The PTI account of observation provides for a synthesis of the long-standing "realism/antirealism" dichotomy, in that both doctrines can be seen as conveying a partial truth. Let us first briefly review these doctrines.

The doctrine of realism spans many forms, from the "naive realism" most of us grow up believing to much more sophisticated forms, including "scientific realism," that have evolved in philosophical debate. For our purposes, we can make do with a definition from the *Stanford Encyclopedia of Philosophy*: "Metaphysically, [scientific] realism is committed to the mind-independent existence of the world investigated by the sciences."[14] The world and the entities in it are assumed by the scientific realist to exist independently of our individual minds, perceptions, and knowledge. The objects in our world are considered as possessing definite properties, which we can come to know without fundamentally disturbing or changing those basic properties.

Antirealism denies this view; it asserts that objects of knowledge are dependent on (or constituted by) some form of subjectivity or mental substance. For example, the philosopher and Irish cleric George Berkeley famously asserted – and ably

---

[13] Subjectivism is the view that knowledge can only be about experiences of a perceiving subject and not about any genuine object external to the subject.
[14] From Chakravartty (2011). I consider only the physical world, not social or political "worlds," for the purposes of this work.

defended – the doctrine *esse est percipi* (to be is to be perceived) and concluded that all objects are ultimately ideas in the mind of God.[15] The work of Immanuel Kant (discussed previously in Chapter 2) is relevant to the realism/antirealism dichotomy because Kant asserted that the only world we can ever come to know is that which depends on the concepts and functions of the human mind: the world of appearance, or what he termed the "phenomenal" realm. Kant did assert that there was "something else out there"; in his terms, the "noumenal" realm, but it was a basic principle of his philosophy that we can never come to know this elusive realm, that which he called the "thing-in-itself." Michael Devitt (1991) refers to Kant as a "weak realist" because Kant did hold that there was *something* that existed independently of our knowledge, even if we could (according to Kant) never obtain knowledge about it.

In the latter twentieth century, Kant's basic approach evolved into a version of antirealism generally known as "constructivism." In Devitt's terms, constructivism asserts that "we make the known world" (Devitt, 1991, p. 236). He correctly (in my view) points out that much of the constructivist argument rests on a conflation of epistemological (knowledge-based), semantic (meaning-based), and ontological (metaphysical) issues. But despite these weaknesses in the usual sorts of arguments for constructivism, it is in quantum theory where this form of antirealism begins to gain traction because of the notorious dependence of property detection on what we choose to measure (recall Section 1.1). In contrast, realism demands that the object of knowledge is *not* fundamentally affected by observation.[16]

We can formulate this dispute in terms of the subject/object distinction presupposed by any discussion about knowledge on the part of an observer (subject) and the aspect of the world the observer wishes to know about (object). In these terms, the realist believes that knowledge is *object-driven*, while the antirealist believes that knowledge is *subject-driven*. We can now make contact with PTI by identifying the "object" with the offer wave and the "subject" with the set of confirmations generated in response to the offer wave components reaching various absorbers. The latter can be thought of in terms of a particular experimental setup or just in terms of the sense organs of an observer.

---

[15] This antirealist doctrine was primarily explicated in Berkeley's *Treatise Concerning the Principles of Human Knowledge* (1710).

[16] The Bohmian theory provides a way to retain realism about quantum objects because it asserts that there really are quantum particles with definite positions, independently of our knowledge or concepts. (Bohmians acknowledge that we disturb those positions in an uncontrollable way when we measure certain contrasting (noncommuting) properties, but that if we choose to measure position, what we find is a particle position that existed independent of our observation. However, I do not favor the Bohmian theory because the "guided particle" ontology is incompatible with the relativistic domain (e.g., recall from Chapter 5 that the classical electromagnetic field must be described by an indefinite number of quanta); there is no account of how guiding waves living in $3N$-dimensional configuration space "guide" particles in three-dimensional space, and its account of the Born Rule depends on taking the wave function as an "equilibrium" distribution in a statistical account, arguably an ad hoc assumption.

Object: Offer wave
Subject: Confirmation wave
Phenomenon: Transaction

Figure 7.5 Subject and object.

With the above identification, PTI can resolve the realism/antirealism conflict by declaring a measured form of victory for both sides. Realism correctly asserts that there truly is "something out there" that is independent of observation. In PTI terms, this is the object represented by a quantum state or offer wave $|\Psi\rangle$. But antirealism correctly asserts that the form the "something" takes is at least partly dependent on how it is observed (in physical RTI terms, detected in an actualized transaction), which takes into account the types of confirmations $\langle\Phi|$ generated by absorbers. Recall from Chapter 4 the man observing the table, reproduced here as Figure 7.5. It's not the "categories" or "concepts" in his mind that do the primary work here, but simply the absorbers in his sense organs. Thus, the "subject/object" dichotomy becomes the "confirmation/offer" complementary relationship in PTI.

The foregoing "defangs" antirealism in the following sense: it need not be anthropocentric, since in PTI, one can have an actual, independently existing event in the absence of a "conscious observer." All one needs is emitters and absorbers, which are physical entities. I should add here that this is not a reversion to physicalism or materialism, since by "physical entity" I mean only an object that is the referent of a physical theory, such as quantum theory. In other words, there is no substance claim pertaining to the term "physical."

This formulation also provides a solution to a long-standing puzzle faced by Kant scholars. The problem is this: Kant insisted that knowledge of the phenomenal world was obtained by way of an interaction of human perceptual activity and concepts with the noumenal world. But the nature of this interaction was deeply obscure. If the noumenal object or "thing-in-itself" was truly "unknowable," what sort of causal power could it have to produce knowledge, even if through human-centered concepts and perceptions? PTI provides at least a partial answer: the noumenal realm is the realm of quantum *res potentiae*; the phenomenal realm begins with the NU-interaction (emission of OW and responding CW), which leads to actualized structured spacetime events constituting the *res extensa* of Descartes. The nature of the interaction between the noumenal realm and the phenomenal realm is just the transactional process.

Thus, in Kantian terms, one can say that the knowable phenomenon is rooted in the "unknowable" noumenon (quantum entity),[17] which gives rise to the transactional process through interactions of OW with responding CW. Actualized transactions result in transfers of energy, which are processed by the senses and their attendant cognitive structures. There are two components to the latter process: (1) physical/ontological (the quantum transaction arising from absorption by the sense organs) and (2) epistemic (the subjective/theoretical concepts used to identify and understand the phenomenon arising from the transaction). The current work deals only with aspect (1) because that is all that is necessary to account for the basic phenomena (the "raw sense data" as described in a Russellian or foundationalist account).[18] As has been noted by other researchers (e.g., Kent, 2010), having to bring in philosophies of mind or explicit psycho-physical dualism weakens the scientific account because there is no account of "mental substance" in the exact science of physics. A traditional "collapse postulate" approach inevitably leads to psychologism of this kind because there is no consistent way to break the linearity of the theory and thereby provide for a determinate result on the physical level without taking absorption into account.

Thus, the transactional model denies the strongest form of realism, namely, the view that objects in their independent entirety are "directly given" to the senses; but it provides support for what is termed "representational realism." The latter assumes that what is directly present to the knower is not the object itself, but "sense data" that make contact with the objectively existing external object and therefore provide authentic knowledge about it. In PTI, sense data are the product of the object, as a source of offer waves, and the subject, as a set of responding absorbers. Together, the subject and object produce transactions that provide information about the object *conditioned on the manner and circumstances under which it is perceived* (really, transacted). The latter sentence is important: such knowledge is always only partial, since transactions vary depending on what types of absorbers are available to the offer waves comprising the object.

---

[17] I put "unknowable" in scare quotes here because I assert that the "noumenal" level can indeed be known, at least indirectly, through quantum theory, whose mathematical formalism tells us something about its structure and dynamics. Thus, I question Kant's assumption that "knowledge" is obtained only through direct sensory observation – knowledge need not be viewed as restricted to the phenomenal realm.

[18] In this regard, I do not deal in this work with the deep and subtle questions concerning the relationship of subjective perception to sense data. However, I do assert that perception properly needs an object, even if not "physical" in the usual sense: perception is transitive and presupposes the fundamental subject–object distinction. (In contrast, one might refer to a perception-free account of experience as *awareness*, which is the ability to perceive.) I assume that *whatever* it is that is subjectively perceived can be attributed to physical transfers of energy via actualized transactions. In cases of nonveridical or hallucinatory perception, an account may be possible in terms of atypical biological processes in the hallucinating subject that ultimately can be traced to transactions among the microscopic constituents of biological components (e.g., neurons).

7.5 An Example: Phenomenon versus Noumenon

This section makes contact with Shakespeare's famous verse that opened this chapter. Let us consider an example of the way in which a *phenomenal* world of appearance, thought of as occurring in "spacetime," arises from a transcendent *noumenal* level in terms of an aspect of popular culture: Internet-based "massive multiplayer online role playing games" (MMORPGs), such as *World of Warcraft* or *Second Life*.

In the game *Second Life*, a player can access an online game environment by loading a software package on their local computer. The player uses the software to create a character, or "avatar," which represents the player in the online game environment. Let's call the human player "Jonathan" and his game avatar "Jon." Once Jon is established in a game environment, he carries with him a point of view (POV) through which Jonathan can perceive what Jon perceives as the latter pursues his in-game career. Now, suppose Jonathan decides to have Jon create something – a table, for example. Jonathan can input certain commands through Jon into the game environment, and a "table" will appear at the desired "location" in Jon's vicinity.

Now, consider another human player, Maria, whose game avatar is "Mia." Maria might be sitting at her computer in Sydney, Australia, while Jonathan is in Montreal, Canada. Nevertheless, their avatars may be in the same game environment "room," say, the "Philosophy Library," where Jonathan/Jon has just created his "table." Now, suppose Jon and Mia don't know that they are only avatars, but assume that they are autonomous beings. We might imagine Jon and Mia discussing the table in front of them along the same lines as the discussion in Bertrand Russell's *The Problems of Philosophy*, chapter 1. For readers unfamiliar with this material, Russell's discussion involves noting that the appearance of the table depends, to a great extent, on the different conditions under which it is viewed (or, more generally, perceived). These appearances may be mutually contradictory: for example, the table may appear smooth and shiny to the eye, but rough and textured under a microscope. Following this line of argument, Russell famously concludes that the only knowledge we can have of the table is of various aspects of its *appearance*, which must always be contingent on the conditions under which it is perceived, and that the "real" table underneath the appearances – whatever that might be – is a deeply mysterious object. In his words:

Thus it becomes evident that the real table, if there is one, is not the same as what we immediately experience by sight or touch or hearing. The real table, if there is one, is not immediately known to us at all, but must be an inference from what is immediately known. Hence, two very difficult questions at once arise; namely, (1) Is there a real table at all? (2) If so, what sort of object can it be? (Russell, 1959, p. 11)

Russell's presentation is an account of the deep divide between, in Kant's terms, the world of appearance (phenomenon) and the thing-in-itself (noumenon). (Notice how he repeats the phrase "if there is one," to emphasize how little we really know about it.)

If Jon and Mia pursue this analysis, they, too, find that the only knowledge they have of the table is based on its appearance (which their human players can monitor on their computer screens showing their avatars' POVs). Suppose the side of the table facing Jon is black and the other side, facing Mia, is white. Jon and Mia can talk to each other and discuss what they see, and they can agree to compare their perceptions by, say, changing places. Then Mia can confirm that the other side of the table is black, and vice versa. By performing these sorts of comparative observations, Mia and Jon can convince themselves that there "really is" a table there because they can *corroborate* their different perceptions in a consistent way: their intersubjective observations form a coherent set. This suggests to them that there is "something out there" that is the direct cause of their perceptions. In commonsense realist fashion, they might conclude that there is a "real" table behind or underneath the appearances – a "table-in-itself"– that "causes and resembles" their perceptions of it.[19]

But what about Jonathan and Maria? They both know that, while the "table-in-itself" could be said to be the cause of Jon's and Mia's perceptions of the game table, the "table-in-itself" *does not "resemble" the game table at all.* What is the "table-in-itself"? It is nothing more than *information* in the form of binary data, manipulated by the people who created the game and by the human users (Jonathan and Maria). Compared with the game table perceived by Jon and Mia, it is insubstantial, abstract. And yet, clearly, *it is the direct cause of the avatars' perceptions of an ordinary table* (the "table-of-appearance") which, to them, is *not* just an "illusion." The avatars cannot ignore it (e.g., they will bump into it and may even incur physical damage if they try to run through it as if it isn't really there). If a human user were to somehow speak to an avatar like Mia and tell her that the objects in her world are nothing but information, she would scoff at the suggestion, and might ask why she suffers damage if she falls off a cliff in her "only information" world. To the avatars, their world is perfectly concrete and consequential.

What does this little parable tell us about our world of "ordinary" objects-of-appearance, that is, our empirical world? It tells us that it is conceivable and even quite possible that the "table-in-itself" of *our* world is a very different entity from what the table-of-appearance might suggest. Because we, and the objects around

---

[19] The naive realist notion that independently existing objects outside the mind are the causes of ideas (perceptions) that resemble them is extensively critiqued in Descartes' *Meditations*.

us, are governed by the laws of physics (the "rules of the game," if you will), we interact with them and are affected by them, and in that sense they are certainly real, just as the game-environment objects are real for Jon and Mia. But the "object-in-itself" is precisely *that aspect of the real object which is not perceived.* If such an aspect exists at all, we can reasonably expect it to be on an entirely different level from our perceived world of experience. Indeed, in terms of PTI, the "object-in-itself" can be considered to be the emitter of offer wave(s) giving rise to possible transactions establishing the appearances of the object. Just as the "table-in-itself" behind the avatars' table does not really live in their game world and is a kind of abstract information, so the quantum entities giving rise to our real empirical objects do not live in spacetime and can be considered a kind of abstract but physically potent information – that is, the physical possibilities first introduced in Chapter 4.

Now, recall from Chapter 2 that Kant asserted that the "thing-in-itself" is *unknowable.* I wish to contest this, based on two main (disparate) points: (1) the fact that Kant has already been shown to have been mistaken in assuming that Euclidean (flat) space is one of the "categories of experience"[20] and (2) the fact that *perceiving* (i.e., sensory perception) is not equivalent to *knowing*, since knowledge can also be obtained by intellectual (rational) means.[21] Concerning (2), recall the arguments in Chapter 2 that an empirically successful principle-type theory can be taken as providing new theoretical referents to previously unknown structural properties of the world. Such an approach to new knowledge is an *intellectual* or rational one rather than an empirical one, the latter being dependent on observation through sensory perception (including the use of sense-enhancing technologies such as microscopes or telescopes), and therefore subject to the limitations of the actualized domain of appearance. In contrast, unexpected but fruitful theoretical development can be considered as pointing to a more abstract (non-observable) level of reality inaccessible to observation, as in the postulation of atoms. The latter was an intellectual step forward in knowledge, not an empirical one.

Recall also that Bohr asserted that the quantum object is something "transcending the frame of space and time" – suggesting (albeit despite himself) an altogether metaphysically new type of entity. The Hilbert space structure of quantum theory greatly exceeds the structure of the empirical world in that it precludes our ability to attribute always-determinate classical properties to objects

---

[20] This could be considered the "Kant's credibility is already suspect" argument.

[21] That this is the case is demonstrated by the great empirical success of physical theories arrived at through rational analysis and mathematical invention. In Einstein's words: "How can it be that mathematics, being after all a product of human thought independent of experience, is so admirably adapted to the objects of reality?" (2010). Nature seems to be inherently mathematical and logical; were that not the case, theoretical science could not provide any useful knowledge.

(recall Chapter 1). Therefore, it's natural to suppose that the structure of the theory describes something "transcending the frame of space and time" but which is nevertheless real because objects described by those Hilbert space states can be created and manipulated through procedures in the laboratory.[22]

We can use the game analogy to immediately gain insight into the phenomenon of "nonlocality." While the avatars and their objects have a maximum speed c, Jonathan and Maria transcend the game environment and can freely communicate instantaneously (with respect to the game environment), so that information can be transmitted from one region in the game environment to any other at infinite speed. This is precisely because that information is *not actually contained in* the game environment. So, for example, Mia might shoot an arrow at game-speed c in Jon's direction while Maria tells Jonathan (over the phone) that she is doing so. Instantly, Jon can step aside and miss the arrow, even though he should not be able to do so according to the rules of the game environment (which would preclude Jon from seeing the arrow coming at him). "Faster-than-light" or "nonlocal" influences are evidence of physically efficacious information existing on a level other than that of the usual local processes (i.e., the game environment or "spacetime").

7.6 Causality

In this section, I consider the vexed notion of "causality" and discuss how transactions can illuminate this long-standing conundrum.

7.6.1 Hume's Elimination of Causality

The reader may recall that the Scottish philosopher David Hume first cast enormous doubt on this commonplace notion of everyday life. As a strict empiricist, he looked for specific evidence of causality in the empirical (observable) world and could not find it. For example, consider a billiard game. The player strikes the cue ball; the cue ball moves and strikes another stationary ball. Subsequently, the second ball moves with the same momentum as the cue ball, which comes to a halt. It is perfect common sense that the cue ball *caused* the second ball to begin moving. However, we never actually *see* the cause; all we see is the pattern of events, which is repeated every time we perform these actions. The reader may object: but surely, we saw the cue ball strike the second ball. How could the second ball *not* move, since it was hit by the cue ball, which we clearly

---

[22] Here I endorse Hacking's dictum that "if you can spray them then they are real" (Hacking, 1983, p. 23), referring to an experimentalist's comment that he could "spray" a piece of equipment with positrons.

observed? But notice again that we did not actually see the cause; the cue ball striking the second ball is not *observably* a "cause." It is simply an event. Our expectation that the second ball must move is based on the fact that we have always seen this happen. It is certainly conceivable that the second ball could just sit there, despite having been hit. The motion of the second ball is predicted by physical law; but again, physical law simply describes patterns of events; it does not say *why* they happen. For this reason, Hume concluded that causation is not really in the world, but is something we *infer* from what he termed the "constant conjunction of events."

Another aspect of the "common sense" of causality (despite the fact that we never actually *see* it) is that the cause always precedes the effect: in terms of the above example, the cue ball striking the second ball *precedes* the motion of the second ball. The contingent, empirical time-asymmetry of causation is addressed further in Chapter 8. For now, I note that this feature of causation is simply a feature of the types of patterns that we see in the empirical world, and should not be thought of as necessarily extendable to the unobservable entities of the micro-world (e.g., electrons), as is customarily assumed.

7.6.2 Russell, Salmon, and Others

As might be expected due to its unobservable nature, the concept of causality is a very slippery and elusive notion. Many distinguished philosophers have attempted to chase it down and capture it in definitive terms, without conclusive success. Bertrand Russell initially expressed great skepticism about causality in this famous quip:

The law of causality, I believe, like much that passes muster among philosophers, is a relic of a bygone age, surviving, like the monarchy, only because it is erroneously supposed to do no harm. (1913, p. 1)

Russell nevertheless felt that causality needed to be well defined in order to support the development of physical laws that seemed to imply causal processes (even if physical laws do not explain them). He developed a theory of causality in terms of "causal lines" (Russell, 1948). This theory was based on several reasonable postulates, such as the idea that there is a kind of "quasi-permanence" in the world: we do not see utter chaos, with objects suddenly and randomly changing their properties. However, Russell's theory was far from bulletproof and came under sustained and cogent criticism from Wesley Salmon (1984), who proposed his own theory of causation. Salmon sought to distinguish genuine causal processes from "pseudo-processes" consisting of effects that are not causal in the usual sense. An example is a moving spot of light on a wall that can exceed the speed of light (see

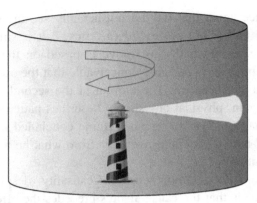

Figure 7.6 The moving spot that "exceeds the speed of light."

Figure 7.6). In that case, no material object actually exceeds the speed of light, but an observable artifact does.

Salmon endeavored to capture the essence of causality in terms of the ability of a causing event to transfer a "mark" to the affected event (some persistent change in the second event which is the effect). However, this theory, too, has been found to have loopholes that hinder its ability to distinguish between what we consider to be genuine causal process and pseudo-processes, such as the moving spot of light or the changing portions of a charged metal plate in shadow (Salmon, 1997, p. 472). Another weakness, according to critics, is that it does not take into account processes such as trajectories of bodies, in which an earlier state seems to serve as the "cause" of a later state. This issue is related to the notorious philosophical riddle of identity and persistence of particular objects. The story of the philosophical pursuit of "causality" as an ontological entity is thus one of the attempts to construct theories of causality that exclude all situations that we regard as noncausal and include only those that we regard as causal.

There has been no conclusive resolution to this puzzle, and I suggest that this is because Hume and Russell were right: causality (at least in its deterministic form) is *not* an ontological feature of the world. In TI terms, it is an inference we make based on situations involving very probable transactions (i.e., transactions with weight close to unity). It can be seen as a supporting feature of physical law because overwhelmingly probable transactions underlie the empirical expression of such fundamental physical principles as the "principle of least action" (recall Section 4.4.1).

7.6.3 Transactions to the Rescue

We can understand the distinction between "causal processes" and pseudo-processes in terms of the transactional process. A transaction constitutes a transfer of energy from the emitter to the absorber. The spot in Figure 7.5 is a locus of absorption and reemission of photons, and in microscopic terms this means that a photon offer wave is annihilated and a new one emitted at that point. Thus the *location* of photon-annihilated emission is moving at a speed greater than c, but no energy is actually being transferred faster than c. We can also account for the apparent persistence of macroscopic physical objects in terms of transactions; recall the baseball of Section 7.4.1, whose persistence as a phenomenal, spacetime object depends on transfers of energy via transactions between its quantum constituents (which comprise the baseball-in-itself) and our sense organs (or any other absorbers). If "earlier states cause later ones," it is in terms of such energy transfers.[23]

Other pseudo-causal processes can similarly be ruled out by reference to the transactional picture. For example, transactions allow us to unambiguously demarcate genuine persistent objects from pseudo-persistent "non-objects," such as the parts of a charged metal plate in shadow, only when they are in shadow (Salmon, 1997, p. 472). Dowe (2000, pp. 98–99) replies that this is not a causal process because the above is not a genuine object – it does not possess identity over time. However, the burden is then on Dowe to define what constitutes identity over time – which he takes as primitive and thereby, according to Psillos (2003, p. 124), makes his account circular. We can define the persistence of an object through time as attributable to ongoing transactions between its constituents and external absorbers, as discussed above in connection with the baseball example. A persistent object also has intramolecular and/or interatomic forces binding its constituents. The charged metal plate as a persistent object in spacetime is a network of transactions whose macroscopic cohesiveness is supported by intermolecular bonds; the changing set of portions of the charged plate in shadow is not. A shadow is just an area of the plate not accessible to an OW from an outside emitter.

7.7 Concerns about Structural Realism

I conclude this chapter by considering a higher-level issue of interpretive methodology. I noted earlier that PTI can be considered in a weaker, structural realist (SR) form which remains agnostic about what these subempirical offer and confirmation waves "really are" in ontological terms. In that regard, I should address some objections to SR, which was first developed by Worrall (1989) in an attempt to circumvent the so-called pessimistic induction concerning the ability of

---

[23] As we'll see in the next chapter, microscopic objects with nonvanishing rest mass can be understood as persisting, in a temporal sense, by reference to their internal clocks defined by internal periodic processes.

scientific theories to refer to ontological entities. The "pessimistic induction" consists in pointing out that many of those supposed entities (e.g., "phlogiston") were later found not to exist; thus, based on past experience, it seems likely that the putative entities referred to by a currently successful theory might also be repudiated. Worrall proposed instead that successful theories refer to *structural* aspects of the world, even if it could not be known what the specific nature of those structures were.

Psillos (1999) has argued that Worrall's distinction between structure and nature cannot be maintained:

> To say what an entity is is to show how this entity is structured: what are its properties, in what relations it stands to other objects, etc. An exhaustive specification of this set of properties and relations leaves nothing left out. Any talk of something else remaining uncaptured when this specification is made is, I think, obscure. I conclude, then, that the "nature" of an entity forms a continuum with its "structure," and that knowing the one involves and entails knowing the other. *(pp. 156–57)*

First, it should be noted that Psillos appears to include the notion of a property as an aspect of structure, when perhaps Worrall has in mind that properties inhere in substance, not in structure. But aside from this ambiguity, we can note the following. The above characterization could be considered as applying to empirical phenomena, perhaps, but not necessarily to subempirical entities. That is, one can consistently propose that the structure of quantum theory dictates that the entities described by the theory cannot be considered to exist within the confines of a spacetime manifold (since the relevant mathematical space for N quanta is $3N$-dimensional and therefore not mathematically commensurate with spacetime). Therefore, we can remain agnostic about the precise nature (i.e., substance) of those entities but still insist, based on empirical success of the theory, that their dynamical *structure* − the way that they behave, including in a relational sense − is accurately captured by the form of the theory. The theory says how the entities are structured but not what they are: in Aristotelian terms, it provides their "formal cause" but not their "material cause" (if any!).[24]

Thus the key difference between the current proposal and typical structural realist proposals is that it denies the usual unexamined identification of "real" with "empirical." For example, Barnum (1990) offers the following comment concerning a formulation by Dieks:

> In Dieks' view, his semantical rule is the sort of thing which is necessary in any attempt to interpret a physical theory: "certain parts of the models [of the theories] are to be identified

---

[24] Aristotle proposed that all objects have four types of cause: material (relating to its substance), formal (relating to its structure), efficient (relating to its creator), and final (relating to its purpose).

as empirical substructures; i.e., part of the theoretical models have to correspond to observable phenomena." I agree with this general characterization of the interpretation of theories: the "internal meaning" of the terms of the theory, given by the mathematical structures which are models of the theory, needs to be supplemented by "empirical meaning." This is done by showing how the theory relates to our experience. (p. 2)

This characterization certainly has had its merits in connection with classical theory, in which all physical entities can be considered as existing in the empirical arena of spacetime. However, the above approach would seem too restrictive for quantum theory, whose structure is incommensurable with that empirical arena. We already know what parts of quantum theory relate to our experience – that is, the probabilities given by the Born Rule – but the point of a realist interpretation of the theory is to go *beyond* that, to find a physical referent for those parts of the theoretical model that cannot be identified as empirical substructures. Thus I agree with Ernan McMullin (1984), who notes that part of the interpretational task is to discover to what the theoretical quantities refer, without assuming that they must refer to something in the macroscopic (empirical) world:

[I]maginability must not be made the test for ontology. The realist claim is that the scientist is discovering the structures of the world; it is not required in addition that these structures be imaginable in the categories of the macroworld. *(p. 14)*

McMullin's point above is a subtle but crucial one, which cannot be overemphasized in connection with the present work. Specifically, my claim is that quantum states refer to something subempirical, yet *real*. As noted previously, this is a new category which is not part of the macroworld, and it is not legitimate to reject it based merely on perceptions that it might seem "implausible" or "unimaginable" when compared with the categories of the macroworld. Of course, it is bound to be counterintuitive since it is not part of our usual empirical experience. Yet one can still show "how the theory relates to our experience" by specifying the conditions (i.e., the actualizing of transactions) under which the subempirical entities give rise to empirical events.

Psillos' objection thus begins with a premise with which I would disagree, namely, "To say what an entity is . . .": a structural realist is not committed to the claim that a theory always says what an entity *is* – that it gives an "exhaustive specification" in usual spacetime or substance/property terms. In fact, this was exactly Newton's interpretive stance when asked to what "gravity" refers.[25] Newton clearly regarded his theory as *about* gravity and as *referring to* gravity;

---

[25] Concerning the ontology of gravity, Newton famously stated "*Hypotheses non fingo*" (I feign no hypotheses); from his *General Scholium* appended to the *Principia* of 1713.

thus, he was realist about his theory. But his theory did not spell out the specific ontological nature of gravity.[26]

PTI in its strong form does go beyond the original TI by proposing a specific ontological referent in the form of physical possibilities. Nevertheless, if one is reluctant to embrace this new metaphysical category, one can still allow that TI captures an essential *structural* element of quantum systems (advanced solutions arising from absorption) missing in the usual account, and thereby provides a more complete interpretation than its competitors.

In the next chapter, I consider the nature of spacetime in PTI/RTI.

---

[26] A similar argument is presented in Dorato and Felline (2011): "we propose, therefore, that the properties of the *explanandum* are constrained by the general properties of the Hilbert model [of quantum theory]. In this sense the *explanandum* [e.g., how or why quantum systems obey Heisenberg's uncertainty principle] is made intelligible *via* its *structural similarities* with its formal representative, the *explanans* [e.g., representability of such systems by Fourier expansions]. Given the typical axioms of quantum mechanics . . . any quantum system exemplifies, or is an instance of, the formal structure of the Hilbert space of square summable functions" (p. 6 in preprint version).

8

RTI and Spacetime

The ontology of spacetime in RTI differs quite radically from the usual conception of spacetime that underlies theorizing in physics. In particular, RTI does not assume the usual notion of a "spacetime background" that is either implied or explicitly invoked in physical theories. In order to get a sense of the distinction, we begin with a qualitative, conceptual discussion, and then turn to a more quantitative account.

8.1 Recalling Plato's Distinction

Let us first recall the philosophy of Plato, discussed briefly in Chapter 1. Plato distinguished two levels of reality: (1) "appearance" and (2) "reality." In Plato's philosophy, (1) means the world as directly perceived by the five senses and (2) means the unperceived world that can be understood by the intellect. In modern terms, these two realms would be called (1) the empirical and (2) the ontological (or extra-empirical) realms, respectively.

Now, the traditional task of physics is to attempt to describe *all* of reality – including that which is not apparent – by accurately observing and insightfully analyzing the world of appearance using logic and mathematics. In other words, physics studies the empirical realm in order to understand *both* the empirical and extra-empirical realms. (A strict empiricist would deny that the job of physics is to gain knowledge of an extra-empirical realm even if it exists. But that approach can be seen as evasion of the scientific mission, as argued in Chapter 2.)

8.1.1 What Is the Empirical Realm?

First, let us consider the question: What exactly *is* the empirical realm in strict physical terms? It is often thought of as "everything in spacetime." However, this cannot be right if the empirical realm is precisely the world of appearance – of

Figure 8.1 The "now" is the direct empirical realm of a particular observer.

direct experience – since we can experience neither the past nor the future (even with powerful instruments).[1] All we can experience is the present, the "now" as it is presented to our senses. So, if we really want to be careful about it, *only the now is the empirical realm.* What do we directly experience about now? That it presents properties to us that are always changing.[2] That is, *the now does not "move"; it changes.* How do we experience these changing properties? We experience them by way of electromagnetic signals that transfer energy from what we are observing to our sense organs (by way of actualized transactions). Thus, our "now" is defined by a spatial coordinate (or, in a relational view of spacetime,[3] the object(s) with which we are currently in direct contact) and any light signals that have reached our eyes from other objects. This is illustrated in Figure 8.1, where the now is symbolized by the person's chair (let's call him "Ty") and the light signals reaching his eyes from objects in his past. While only Ty's "now" is his *direct* empirical realm in the sense of immediate experience, the messenger photons bring him information about certain objects in his past, so we can include those past objects as empirically available to him in that respect. Also, the messenger photons themselves experience neither passage of time nor spatial displacement, which (from the point of view of the photons) puts Ty in a kind of direct contact with those objects, even though there is temporal and spatial displacement from his point of view.

This means that, strictly speaking, each observer has their own, unique, "empirical realm." However, we can corroborate our experiences and arrive at a consistent *intersubjective* consensus about a "larger" world of appearance beyond

---

[1] This holds for observations of distant astronomical objects, as follows. The light we detect from a galaxy 10,000 light years away left that galaxy 10,000 years ago, but we don't actually see it until it reaches us in the present. Thus, we see the galaxy in the present as it was 10,000 years ago (as measured in our inertial frame, of course). We don't actually experience the past. This is the same as getting a message in a bottle from a castaway. The castaway may be long dead, but the message is something written while the castaway was alive.

[2] Norton (2010) makes this point as well: "we do have a direct perception of the changing of the present moment. That is clearest in our perception of motion."

[3] I discuss relationalism below. In a nutshell, relationalism denies that spacetime exists as an independent substance or "container."

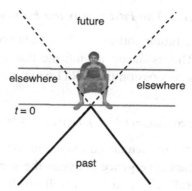

Figure 8.2 A spacetime diagram.

our individual empirical realms. All of these corroborations are conducted using photons, by way of transactions that establish the emission and absorption events for the transferred photons. It is in this sense that the "fabric of spacetime," which according to RTI is constructed of transacted photons and their emission/absorption events, can be thought of as a collective empirical realm.

The above reinforces the idea that relativity theory (with its limitation of signal speeds to the speed of light) places restrictions on the empirical realm. But the empirical realm is in fact even more restricted than is often noticed in discussions of relativity. For example, consider a typical spacetime diagram such as Figure 8.2. Besides the diagonal "light cone" lines, the diagram indicates gray horizontal "lines of simultaneity" for a given reference frame. But even though these lines (really three-dimensional hyperplanes) are defined with reference to a given observer (Ty), almost all of the points comprising those lines are extra-empirical: they are not within Ty's empirical realm. For example, the hyperplane $t = 0$ crossing Ty's "now" is extra-empirical (note that it is in the "elsewhere" region) except for those points in direct contact with him. Owing to the fact that electromagnetic signals have a finite speed, no observer *really* sees or touches anything outside the apex of their light cone.[4] When you sit in a chair reading this book, you are seeing the page as it existed a few nanoseconds ago, *not* as it exists along a line of simultaneity from your eyes.

---

[4] It may seem lonely to realize that each of us sits isolated atop the apex of our light cone. But recall Black Elk's vision: "'I saw myself on the central mountain of the world, the highest place, and I had a vision because I was seeing in the sacred manner of the world.' ... And then he says, 'But the central mountain is everywhere'" (Neihardt, 1972). This can just as easily describe the way each observer is at the top of their own empirical "mountain." It is also strikingly reminiscent of the way the South American indigenous people the Aymaras view time: in contrast to the typical Western view, they see themselves as "facing the past" with the future behind them (as unknown/unknowable).

8.1.2 The Past versus the Future

Note in Figure 8.2 that the future portions of Ty's light cone are dashed, while the past portions are solid. This is meant to indicate that in RTI, the future is not actualized but exists only as possibilities.[5] RTI exemplifies Hans Reichenbach's dictum:

The flow of *time* is a real becoming in which *potentiality* is transformed into actuality.[6]

The assertion that there are no actualized events in the future light cone of any observer marks a significant divergence between the spacetime ontology of RTI and the usual "block world" view, so let's dwell on that for a moment. What it means is that, while we could certainly draw another light cone centered on a hypothetical observer in the future light cone of Ty, that image would be merely a conceivable possibility (category III in Figure 7.2) which does not correspond to our world.[7]

This account of the future as unactualized possibilities is based on the following considerations. In the RTI formulation, spacetime events (as end points of invariant intervals defined by real photon transfers) are established only through actualized transactions, and therefore *such events do not exist prior to transactions*. This means that just prior to the Big Bang, which is identified by the onset of transactions, *there was no spacetime*. Prior to the Big Bang, all that existed was quantum possibilities in a primal "now." The "fabric of spacetime" is created in the present and recedes from the present into the past (see Section 8.1.3 for elaboration). So, at the moment of the Big Bang, spacetime itself began to be created. The expansion of the universe is not just the aftermath of an "explosion" in the classical sense; rather, it corresponds to the continual creation of spacetime intervals through actualized transactions, a process that began with the Big Bang.[8]

Since, according to the relational spacetime ontology proposed here, the "fabric of spacetime" is no more than the structured set of actualized events in the ever-receding past, there can be nothing actualized in any observer's future light cone – including, of course, another hypothetical observer. In metaphorical terms, "spacetime" is the cast-off skin of a snake; the living snake is eternally in the present. The structure of spacetime, as described by relativity theory, is the map that allows different observers (more precisely, any system with nonvanishing rest mass) to coordinate their information about the snakeskin(s) in a consistent way.

---

[5] This basic picture of time is termed "possibilism" in Savitt (2008), although the present model differs from the usual "growing universe" or "possibilist" temporal theory, as will become apparent.

[6] Reichenbach (1953).

[7] This is the usage of "possibility" corresponding to the nonphysical, unreal possibilities of category III in Chapter 7.

[8] This has interesting implications for the origins of both "dark matter" and "dark energy"; see Kastner et al. (2018).

Thus the spacetime diagram, because it is so easily subject to arbitrary event placements in a hypothetical "future," typically misleads us into thinking that there can be "future events" and "future observers" when – according to the model proposed herein – this is physically not the case. Just because we can draw something on a spacetime diagram does not mean that it can physically exist in our world. The notion that the ability to represent something on a spacetime diagram implies that it may physically exist can be very compelling.[9] However, to see why we need to be wary of subscribing to this unwarranted assumption, consider the following analogy. Animation artists now have programs that can do a lot of the tedious work of redrawing frame after frame of the same character for them. A typical animation program allows you to load a basic image of a character, indicating where all the joints are, and the program will change the angles of the joints for you in a series of images to make the character appear to move. You have to specify the amounts by which each of the joints is to move in each frame. Theoretically, for example, you could make a character's head turn by any amount in any direction, but that doesn't mean that the motion will be realistic or even physically possible. The spacetime diagram similarly lacks certain physically relevant constraints. So the freedom to draw whatever we choose, wherever we choose, on such a diagram does not imply that what we drew corresponds to what is physically possible, any more than the freedom to make a character's head spin around in circles in an animation program means that this would be possible in the real world. As observed in Chapter 2, the map (i.e., the spacetime diagram) is *not* the territory and can correctly represent only certain specific aspects of the territory.

8.1.3 The Fabric of Created Events

To gain further insight into the proposed spacetime ontology, consider the following metaphor. Think of the past as a knitted fabric (see Figure 8.3). The present is the stitch currently being knit, whatever the time index t (here, indexing the row) of that stitch. Let's assume that $t = 0$ corresponds to the Big Bang, when the first stitch is "cast on" to the needle. The future is nothing more than one or more balls of yarn of different types, a pattern, and/or some ideas about what to knit. The present or "Now" is the realm in which our garment is created. The Now doesn't "move," but the stitches on the needles change (perhaps in color or texture) and are extruded away from the Now in the form of fabric as the knitting progresses. The creation of a new stitch is always attributable to the actualization

---

[9] This is basically the same point, albeit with respect to "spacetime" rather than just time, made in Norton (2010) in a slightly different context: "We start to get used to the idea that our theories of space and time are telling us all that can be said about time objectively" (p. 26).

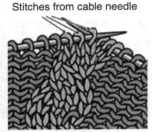

Stitches from cable needle

a. Slip the cable stitches to   b. Knit 3 from the LH needle.   c. Knit the stitches from the
the cable needle and hold   cable needle.
in front.

Figure 8.3 The past as a knitted fabric; "Now" corresponds to the stitch currently being knit on the needles. This picture is a version of C. D. Broad's "growing universe" theory of time (Broad, 1923). It has much in common with the theory of Tooley (1997) in that a statement like "the stitches in row *t* are currently on the needle" can be seen as equivalent to "event E is present at time *t*" in Tooley's picture. This is because, for the former statement to be true when uttered, there is not yet a row $t + 1$ (i.e., the future is not actual). However, I obviously do not adopt other features of Tooley's theory, such as its spacetime substantivalism. Image from www.dummies.com/how-to/content/knitting-cables.html. From Pam Allen, Tracy Barr, and Shannon Okey, *Knitting for Dummies*, 2nd ed. © 2008. Reprinted with permission of John Wiley & Sons, Inc.

of a specific transaction, as discussed above. Thus, the Now is not something that "moves forward"; rather, the Now is the empirically always-present field of change represented by the knitting needle, while *the past is something that continually falls away from us*.[10] So, a locution such as "when $t = 5$ is now" means the stage of the knitting process at which the stitches time-indexed by row number five are on the needle (in more familiar terms, "are in the present"). As the process continues, those stitches are extruded and are no longer on the needle, but they keep their time index as they recede.[11] Thus, if we are knitting a scarf, the Big Bang indexed by $t = 0$ is the bottom edge of the scarf.

The domain of classical phenomena can be characterized as a "fabric" in which the stitches are very small, uniform, and tight, and we can think of the classical laws of motion as predictable colored patterns in the resulting fabric. But, if we "zoom in" on the same fabric (as in the Chapter 7 discussion of "zooming in on a baseball"), that is, deal with a scale at which quantum phenomena are possible, some stitches are metaphorically removed from the needles, giving rise to patterns of a different character (such as cables that seem to "float" above the background

[10] I owe this insight to my daughter Wendy Hagelgans.
[11] These "knitting stages" can be thought of as Stapp's "process time" (Stapp, 2011), though he views the "now" as advancing while I differ from that aspect of the picture.

of knitted fabric). These are quantum phenomena, arising, for example, in the delayed choice experiment (recall Chapter 6).[12]

In Figure 8.3, (a) some stitches are removed from the knitting process and held in an "indeterminate state" (on the cable needle) as (b) surrounding stitches are knitted into the "past" (the extruded fabric). Thus the standard "classical" evolution of the various phenomena continues, except for those "indeterminate" stitches that are held back until a later stage. In (c), the indeterminate stitches are made determinate as they take their place in the fabric. The result is a pattern with more texture and depth than the plain "classical" fabric. As noted above, this is a kind of "growing universe" theory of time;[13] but it is the *past* that grows and continues to become actualized as it falls away from the present – the present does not "advance." Meanwhile, as noted above, the future is not a realm of determinate events but rather a realm of physical possibilities – the "raw material" for events, if you will. The future is a set of possibilities that becomes woven into the created past through actualized transactions.

The foregoing picture is roughly reminiscent of McTaggart's so-called A-series of time, although it has some important differences. To review this terminology: in a famous paper, McTaggart (1908) offered an argument against the independent existence of time by asserting that temporal events need to be characterized in two different (A vs. B) ways. In the "A-series," an event is characterized by whether it is in the past, present, or future; while in the B-series, the same event is characterized only by the usual temporal index t, which allows only "before"- and "after"-type relations between events. McTaggart argued that both characteristics are necessary for time to exist. But, he argued, any given event E indexed by t will "at different times" be past, present, *and* future. In order to specify "when" to apply those differing A-series descriptions, it then appears necessary to invoke an additional time index, say, s. Then, for example, we can truthfully say "event E at time t is in the present when time s is the present." But in order to say when time s is present, we need a third index, and so on, ad infinitum. So, according to McTaggart, statements involving truly temporal properties (e.g., past, present, and

---

[12] I am glossing over some nuance here, since criteria for "quantum" versus "classical" phenomena can be characterized in different ways. Some of these are: entanglement versus nonentanglement, indeterminism versus determinism, wave/particle duality versus particle *or* wave. Here I focus on the indeterministic temporal aspects of quantum-level behavior.

[13] As mentioned earlier, the first "growing universe" approach to time was proposed by C. D. Broad (1923). Earman (2008) gives a critical discussion of Broad-type "growing universe" theories. I believe that the transactional model resolves many of the challenges Earman raises for Broad-type theories. For instance: (1) a direction for "becoming" is clearly specified in terms of the distinction between emission and absorption, and (2) transactions provide the kind of dynamic creation of events that he worries seems to be missing in Broad's original approach. Note that the present model has no problem with future-tensed sentences as outlined in his discussion; the truth value of a future-tensed sentence is indeterminate at the time it is uttered. This is because the model does not consist of a set of ordered "chips off a Newtonian block"; there is genuine indeterminacy in the becoming.

future) cannot be unambiguously true or false (because of the infinite regress involved in attempting to pin down their truth or falsity). He concludes, based on the indefinite truth character of statements about time, that there is really no such thing as temporality in the "passage" or "becoming" sense.

Note, however, that underlying McTaggart's proof is the crucial assumption that the future has the same essential character as the present or the past: specifically, that past, present, and future all purportedly consist of unique, determinate events. This is essentially the block world assumption, which precludes any genuine becoming: it is a static ontology. Genuine becoming of the kind inherent in RTI means that specific, unique events are coming into being (in RTI terms, being actualized) that *did not exist before*. Thus, genuine becoming implies that the "future" is not populated by uniquely established, determinate events, as assumed by McTaggart. If this crucial assumption is rejected, McTaggart's proof cannot go through.

McTaggart's argument has also been presented not in terms of events but in terms of time indices themselves, so that a purported A-series description would characterize a time index such as t_0 as "future," "present," or "past," at different "times" (which clearly leads to an infinite regress). But as we will see in more detail in the next section, according to RTI, these time indices apply to internal quantum periodicities, at the level of the quantum substratum. Thus they are not part of the spacetime manifold and are therefore not subject to characterization in terms of "past," "present," or "future" where these are understood as describing actualized events. While time indices can of course be used to tag actualized events, *it is the events themselves*, not time indices, that qualify as spacetime entities. And again, there are no "future events," since events do not exist until they are actualized, and the present is the domain of actualization.

So, to recap: McTaggart's argument against the independent existence of time does not apply to RTI, for the following reasons:

(1) In RTI, there are no "future times or events," since the future is not actualized (RTI denies the block world ontology implied by McTaggart's definitions of his A- and B-series).
(2) RTI does not depend on some ill-defined notion of "passage" that needs to be characterized by some additional temporal index. Instead, the primary temporal feature is *change*, which characterizes the Now (via changes in the states of transacting quantum systems). So change is primary, and the time index is just a way of recording those changes as the fabric of spacetime falls away from the Now, through actualized transactions.

But more importantly and in general terms, McTaggart's argument could arguably be said to be circular, since its very definitions presuppose that there is no ontological difference between past, present, and future. That is the essence of a block world ontology – a static picture – so temporality in a "becoming" sense cannot apply to it, by definition. Any ontology that has true becoming, in which the future is unactualized and does not consist of unique, determinate events, is therefore immune to McTaggart's argument. Specifically, in a true becoming ontology, there is never a "future event," because before an event is actualized, it does not exist (the B-series is denied). In the case of RTI, the only sense in which the term "future" could be applicable would be as a description of unactualized possibilities, as opposed to specific events (the latter being specific *activities* of quantum systems, as we will discuss below). The present is the domain in which one of those possibilities is actualized and the others are not. Even though one can have pseudo-classical situations in which a particular event is overwhelming likely to be actualized, that event does not exist as a spacetime entity "in the future," since there are no events in the future.

8.1.4 Becoming and Relativity

The above account may raise the worry that a kind of "absolute simultaneity" is being smuggled in, which is at odds with relativity's banishment of that notion. That is, doesn't a "row of stitches" count as a set of simultaneous events? Yes, but only with respect to a given observer, and that does not translate into the claim that events carrying the same time index share the same "Now." As Stein notes, "'a time coordinate' is not time" in relativity theory (Stein, 1968, p. 16). Recall that an event outside an observer's light cone is strongly extra-empirical. Therefore, as Stein (1968, 1991) argues, it is at best physically vacuous, and arguably inconsistent with relativity theory, to attribute "nowness" to an event outside the light cone. Indeed, it should be noted that the knitting analogy is only a partial one in that according to RTI, becoming does not occur in a "row," or along a given plane of simultaneity.[14] Rather, it occurs with respect to invariant spacetime intervals. This issue will be discussed in further detail in Section 8.2.

Of course, it is often claimed that relativity is incompatible with a "becoming" picture of events,[15] but such arguments have depended on (1) a substantivalist notion of spacetime which takes "events" as mathematical points in a preexisting spatiotemporal "substance," and/or (2) an assumption that the "present" is defined with respect to a particular plane of simultaneity, which I have already disputed above. Stein (1968, 1991) makes a persuasive case against assumption (2), even as he uses a substantivalist picture to argue that no event can be (in my terminology)

---

[14] This feature is a key distinction between the current proposal and that of Tooley (1997); the latter posits an absolute space and therefore an absolute rest frame, so Tooley's "fabric" does have a horizontal leading edge.
[15] Cf. Rietdijk (1966), Putnam (1967), Penrose (1989). These are referred to as "chronogeometrical fatalism" by Savitt (2008).

actualized with respect to another event that is spacelike separated from it unless all events are actualized, leading to a "block world."

Under RTI, the fundamental structural component of becoming is the actualized transaction, which establishes only two spacetime "points." (This feature is quantitatively elaborated in Section 8.2.) In this non-substantivalist picture, one cannot use Stein's argument (Stein, 1991, pp. 148–49) that given two spacelike separated events *a* and *b* such that *b* is actual with respect to *a*, all other spacetime "events"[16] must also be actual with respect to *a*, including those in the future with respect to *a*. That argument requires that spacetime be a mathematical manifold of not necessarily occupied points, which I deny.[17] Recall also that the actualization of a transaction is an aspatiotemporal process; it is the *coming into being* of an entire spacetime interval. *The actualization of a given transaction defines the "Now" for the associated absorption event.* It is not appropriate to consider "Now" as applying to an entire manifold of events "at the same time," since the latter phrase smuggles in an inappropriate simultaneity notion.

With regard to an observer's subjective perception of "now," the "present" or the "now" is always a local phenomenon. Significantly, the French word for "now" is *maintenant*, literally, "holding in the hand." The elusive nature of "now" comes from the fact that it is necessarily a noncollective property; it applies to each individual transaction's aspatiotemporal actualization process. In Stein's terms, an event is only present to itself: "in [relativity theory], the present tense can *never* be applied correctly to foreign objects" (Stein, 1968, p. 15). He goes on to express a view of becoming that applies in essential terms to the present model (with the modification that instead of the "spacetime point" referred to below, the "chronological perspective" is that of a system with finite rest mass experiencing an actualized absorption event):

In the context of special relativity, therefore, we cannot think of temporal evolution as the development of the world in time, but have to consider instead ... the more complicated structure constituted by ... the "chronological perspective" of each space-time point. (p. 16)

---

[16] "Events" in quotation marks refers to "unoccupied" spacetime points in a substantivalist approach.

[17] The argument that I critique here is a version of "chronogeometrical fatalism." My version of Stein's "*Rab*" relation, which says "*b* has become with respect to *a*," therefore need not limit *b* to the past light cone of *a*. My model would also appear to be immune to a similar argument of Weingard (1972), since his is also based on a substantivalist view of spacetime and the assumption, based on the conventionality of simultaneity (i.e., choice of the one-way speed of light) that *any* "event" outside the light cone of an actualized event must correspond to an actualized event. I go in the opposite direction, in a sense: one cannot assume that any spatiotemporal index outside one's light cone corresponds to an actualized event. Such an index corresponds to an actual event *only* if it is the absorption or emission site of an actualized transaction, not by reference to the structural features of a preexisting spacetime substance (which I deny). Along with this would go the requirement that an actualized event not be in the future light cone of any other actualized event. But since the set of actualized events is contingent on the actualization of specific transactions, not the structure of a spacetime substance, there is no inconsistency.

Another account of the locality of the empirical "now" is given in Norton (2010):

The "now" we experience is purely local in space. It is limited to that tiny part of the world that is immediately sensed by us. There is a common presumption of a present moment that extends from here to the moon and on to the stars. That there is such a thing is a natural supposition, but it is speculation. The more we learn of the physics of space and time, the less credible it becomes. For present purposes, the essential point is that the local passage of time is quite distinct from the notion of a spatially extended now. The former figures prominently in our experience; the latter figures prominently in groundless speculation. (p. 24)

In the present model, rather than a "passage of time," we have the generation of an ever-increasing "fabric" of past events,[18] but the basic observation is the same: "Now" is a local phenomenon.

8.1.5 The "Dead Past"

Here I address another issue that arises in contemporary discussions of "growing universe" pictures of time, namely, how to understand the "dead past" feature of the model I propose here. That is, "people in the past" (such as Socrates) are not observers having empirical experiences. As noted above, the actualized past is like the cast-off skin of a snake; the living, experiencing snake is no longer contained in it.

This model is in contrast to "presentism," the view that only the present exists. Heathwood (2005) argues, in response to Forrest (2004), whose model is similar to this one, that regarding people in the past as nonconscious leads to the same problems plaguing "presentist" accounts of time. The problem for presentism is that there seems to be no plausible way to account for the meaningfulness of a statement such as "I admire Socrates" if Socrates, being in the past, does not exist. To what, then, does the sentence refer? The growing universe approach sidesteps this problem, since in that approach, Socrates *does* exist in the past. However, Heathwood argues that the same problem reappears in the "dead past" version of the growing universe for sentences such as "Socrates was conscious when he was killed." He certainly has a valid point if such sentences are taken as referring to the "dead past." But I would argue that the appropriate referents of such sentences are just earlier stages in the process of the growing universe. So the referent for the above sentence is the stage at which Socrates' execution was "on the knitting needle," or "in the present" (or, taking into account the previous section, when

---

[18] But see Section 8.2.4 for a discussion of an "internal clock" that applies to all systems with finite rest mass.

Socrates was "present to himself").[19] I see no reason why such statements cannot refer to an earlier stage in the process; language need not be restricted to any particular stage of a growing universe.

8.2 Transactions and Spacetime Emergence

In this section, we will consider in more quantitative terms the manner in which the spacetime manifold emerges from the quantum level by way of actualized transactions, in a process of genuine "becoming" alluded to by Reichenbach's statement above. As noted in the previous section, according to RTI, there is no a priori "spacetime background" as physicists generally assume. Instead, the structured set of events that constitutes the spacetime manifold emerges from the extra-spatiotemporal *quantum substratum* comprising physical *potentiae*, that is, entities described by quantum states. This domain is characterized by Hilbert space structures and processes (such as unitary interactions).

We must first recall, from Chapters 5 and 6, that the only kind of quantum system that is transferred directly from an emitter to an absorber in an actualized transaction is the massless gauge boson, or photon.[20] Quantum systems with nonvanishing rest mass act as emitters and absorbers of photons and are not themselves transferred in transactions, that is, via offers and confirmations. While they can be transferred from one system to another, such transfers involve liberation from one bound state and reintegration into another bound state, a process that is physically distinct from the emission and absorption of a photon in a transaction. In what follows, we will see that the role of emitters and absorbers in spacetime emergence is quite different from that of the photon. We'll also see how transactions create a discrete, interlocking, structured set of events that fulfills the role of "spacetime" without the necessity of invoking a background spacetime "substance" or container. Thus, according to RTI, the usual notion of a "spacetime continuum" is a fiction, as is the notion of a "spacetime background" for all physical processes.

It turns out that a natural way to formulate the process of spacetime emergence is in terms of the concept of a *causal set*. We will begin by reviewing this concept.

---

[19] The same referent would apply to the sentence viewed as unproblematic by Heathwood, "Socrates was fat when he was killed." This consideration thus resolves the concern he raises about inconsistency of the "truth-makers" for the two types of statements.

[20] Thus far, gluons are considered massless, but their status in this regard is uncertain. In any case they are never free particles in the sense that they do not carry radiative energy, so we disregard them as far as transactions are concerned.

8.2.1 Causal Set Approach

A causal set is a partially ordered set that can enlarge in a directed manner. The primary motivation for the causal set program was to solve the problem of quantum gravity. Its originator, Raphael Sorkin, remarked:

The causal set idea is, in essence, nothing more than an attempt to combine the twin ideas of discreteness and order to produce a structure on which a theory of quantum gravity can be based. That such a step was almost inevitable is indicated by the fact that very similar formulations were put forward independently in [G. 't Hooft (1979), J. Myrheim (1978) and L. Bombelli et al. (1987)], after having been adumbrated in [D. Finkelstein (1969)]. The insight underlying these proposals is that, in passing from the continuous to the discrete, one actually gains certain information, because "volume" can now be assessed (as Riemann said) by counting; and with both order and volume information present, we have enough to recover geometry. (Sorkin, 2003, p. 5)

While the transactional interpretation is not a theory of quantum gravity, it dovetails very naturally with the above program in that the structures that emerge from the transactional process feature both discreteness and order, and effectively form a causal set.[21]

In formal terms, a causal set C is a finite, partially ordered set whose elements are subject to a binary relation $\prec$ that can be understood as precedence; the element on the left precedes that on the right. It has the following properties:

(1) transitivity: $(\forall x, y, z \in C)(x \prec y \prec z \Rightarrow x \prec z)$
(2) irreflexivity: $(\forall x \in C)(x \sim \prec x)$
(3) local finiteness: $(\forall x, z \in C)$ (cardinality $\{y \in C | x \prec y \prec z\} < \infty$).

Properties (1) and (2) assure that the set is acyclic, while (3) assures that the set is discrete. These properties yield a directed structure that corresponds well to temporal becoming, which Sorkin describes as follows:

the relationship $x \prec y$ ... is variously described by saying that x precedes y, that x is an ancestor of y, that y is a descendant of x, or that x lies to the past of y (or y to the future of x). Similarly, if x is an immediate ancestor of y (meaning that there exists no intervening z such that $x \prec z \prec y$) then one says that x is a parent of y, or y a child of x, ... or that $x \prec y$ is a link. (Sorkin, 2003, p. 7)

In Sorkin's construct, one can then have a totally ordered subset of connected links (as defined above), constituting a *chain*. In the transactional process, we naturally get a parent–child relationship with every transaction, which defines a link. Each actualized transaction establishes three things: the emission event E, the absorption

---

[21] Nevertheless, the ability of RTI to provide an account of the emergence of elements of the causal set positions it as a useful component of a theory of quantum gravity of the sort Sorkin et al. are exploring.

event *A*, and the invariant interval $I(E, A)$ between them, which is defined by the transferred photon. Thus, the interval $I(E, A)$ corresponds to a link. Since it is a photon that is transferred, every actualized transaction establishes a null interval, that is, $ds^2 = c^2t^2 - r^2 = 0$. The emission event *E* is the parent of the absorption event *A* (and *A* is the child of *E*).

A major advantage of the causal set approach as proposed by Sorkin and collaborators (e.g., Bombelli et al., 1987) is that it provides a fully covariant model of a growing spacetime. It is thus a counterexample to the usual claim (mentioned in the previous section) that a growing spacetime must violate Lorentz covariance. Specifically, Sorkin shows that if the events are added in a Poissonian manner, then no preferred frame emerges, and covariance is preserved (Sorkin, 2003, p. 9).

In RTI, events are naturally added in a Poissonian manner, because transactions are fundamentally governed by decay rates (Kastner and Cramer, 2018). As discussed in Chapter 5, the elementary probability of the NU-interaction (occurrence of a transaction) corresponds to the fine structure constant α. But, owing to conservation requirements, the full expression for the probability of a transaction is essentially the transition probability between emitter/absorber states, *X* (excited) and *G* (unexcited): $|\langle X, 0|H_{\text{int}}|G, k\rangle|^2$. Here, H_{int} is the interaction Hamiltonian quantifying the coupling between the emitting/absorbing charges and the electromagnetic field: $H_{\text{int}} = e\hat{A} \cdot \vec{p}$ in natural units.[22] *k* is the state of the photon that is transferred in order to satisfy conservation requirements.

The squared form emerges in the transactional picture because both emission and absorption are necessary for transfer of the photon, and the photon emission and absorption amplitudes are complex conjugates of one another. When one multiplies both amplitudes together for the complete process, one therefore gets the Born probability of a photon being transferred between these two states. In the squaring of a transition amplitude containing H_{int} we see the origin of the factor of $a = e^2$, the fine structure constant. Applying time-dependent perturbation theory to the specific initial and final states, given the perturbation H_{int}, leads to the standard decay rate, a Poissonian process.[23] Thus, the emergent spacetime structure in RTI is fully covariant. However, it's important to note that, while the original causal set model assumes that individual events constitute the basic volume element of spacetime, in RTI it is the invariant spacetime interval $I(E, A)$ that constitutes the basic volume element. So, rather than a structure that is growing by single events,

---

[22] The direct-action theory can work with the Hamiltonian form because of the equivalence of the traditional quantum field $\hat{A}$ with a direct connection between currents, as discussed in Chapter 5.

[23] Here, the perturbation applies to the interacting fermions, which evolve according to the time-dependent Schrödinger equation. The squaring, strictly speaking, really applies to the photon, which is not tied to any spatiotemporal index (unlike the emitter and absorber for whom inertial frames can be defined). This is related to the issue, discussed in Chapter 6, wherein emitters and absorbers are detected only indirectly by way of photon transactions, and are not themselves transferred via transactions.

the RTI spacetime structure grows by pairs of connected photon emission and absorption events – that is, by links. As noted above, this means that the spacetime manifold itself is really constituted solely of null intervals.

8.2.2 Rest Mass Remains in the Quantum Substratum

Before studying the resulting structure further, we must be clear about an unfamiliar aspect of the proposed ontology. This ontology departs sharply from the usual Democritan concept of "atoms in the void" upon which physics has been traditionally based (however unconsciously at times). That is, it is usually tacitly assumed that physics deals with chunks of something called "matter" moving around in an otherwise empty spacetime container. Matter, in this picture, is an undefined primitive with only operational properties.

However, in the RTI ontology, systems with rest mass, such as atoms and molecules – that is, emitters or absorbers – are *not* part of the spacetime manifold. They remain in the extra-spatiotemporal domain (quantum substratum) described by Hilbert space, even as they undergo state changes as a result of their participation in transactions. This means that the notion of "change" applies just as well to the quantum substratum as it does to the spacetime manifold. In Section 8.2.4, we'll see that quantum-level change can be described by reference to an internal "clock" of quantum systems with finite rest mass.

Emission and absorption *events* such as E and A above must be distinguished from the emitter and absorbers themselves; an event is *not* a rest-mass quantum system. An event is not an entity or substance; rather, it is an *activity* of an entity. The only elements of spacetime are emission and absorption events and the real photon defining and connecting the two events. Thus, emission/absorption events and the transferred photons (constituting links) are aspects of spacetime structure; everything else is not, and abides in the quantum substratum.[24] This substratum is the source of the emerging, growing spacetime structure, much as the mineral-laden water is the source of the crystals in a geode (recall Chapter 4). The disanalogy here is that rest-mass quanta do not themselves transform into spacetime objects; only the electromagnetic field does so, in the form of photons that serve as structural elements (links) of the emergent spacetime manifold.

---

[24] Besides nonlocality and entanglement, another aspect of the departure of quantum-level processes from long-standing empirical-level physical principles is found in the fact, as noted in Brown (2005), that quantum test particles do not obey the "zeroth law of mechanics," that is, the principle that "the behavior of free bodies does not depend on their mass and internal composition" (Brown, 2005, p. 25). This is easily seen by looking at the time-dependent Schrödinger equation, which depends explicitly on the mass of the quantum. Again, this discrepancy can be understood by considering quantum mechanics as describing the behavior of subempirical (pre-spacetime) objects that do not have to obey empirical-level principles of mechanics.

This picture has much in common with Ellis and Rothman's "crystallizing block universe" (CBU), which is a particular sort of "growing spacetime" (Ellis and Rothman, 2010). However, the CBU ontology seems to assume a spacetime background, even if the future is taken as indefinite. Thus, RTI differs somewhat from the CBU picture in that in RTI the quantum formalism specifically refers to an extra-spatiotemporal domain, or quantum substratum, from which the spacetime structure emerges. In addition, the actualization of events in RTI does not correspond to a "moving present" that progresses "toward the future" as is the case in the CBU. Instead, as discussed in the previous section, the generated spacetime structure *recedes* from a present that is eternal in some sense, since it is just the interface between the quantum level of possibilities and the growing spacetime manifold. Again, the knitting analogy can be helpful here, although it should be kept in mind that there is no preferred reference frame corresponding to a "knitting needle." In this respect, it agrees with the CBU picture in that it is interacting matter and energy that locally generate actualized events, in a manner that does not single out any preferred reference frame.

In addition, under RTI, the actualization of measurement results corresponds to a specific quantitative physical process (i.e., the transactional process or NU-interaction), and it is not dependent on decoherence arguments or top-down considerations, as is the case with the CBU. Rather, the actualization of spacetime events and the attendant arrow of time emerge from the micro-level, through the non-unitarity of the NU-interaction. We will consider this issue in further detail in Section 8.3.

8.2.3 The Basic Structure of the Emergent Spacetime Manifold

With the above in mind, let us recall the parent–child relationship introduced in the previous section, and consider a single actualized transaction involving an emitter C, and its receiving absorber, D (refer to Figure 8.4). C and D are bound systems such as atoms or molecules, that is, systems with internal degrees of freedom subject to excitation. D is the absorber that actually receives the real photon as a result of the final collapse (or reduction).

The atoms' initial roles as emitter and absorber can be represented by denoting their initial states as $|X\rangle_C$ and $|G\rangle_D$, respectively, where X is the excited state and G the ground state. Let us designate the initial emission event as E_1. This event heralds C's transition from the excited state to the ground state, $|X\rangle_C \rightarrow |G\rangle_C$. Similarly, the absorption event A_1 heralds D's transition from the ground state to the excited state, $|G\rangle_D \rightarrow |X\rangle_D$. The newly created link is a null interval established by the transferred photon (indicated by the wavy line) and bounded by E_1 and A_1: symbolically, $I(E_1, A_1)$. The figure schematically depicts the idea that

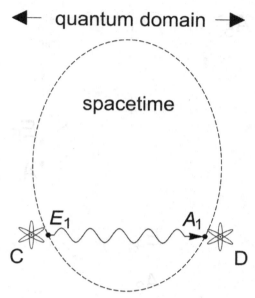

Figure 8.4 A single transactional link is created.

the emitter and absorber remain beyond the spacetime construct, in the quantum substratum, while the exchanged photon establishes a link that constitutes an element of spacetime. In addition, the bold dots indicate the spacetime events of the emission and absorption (again, these events are not identified with the emitting and absorbing systems but rather are activities of those systems).

Following the absorption event A_1, D is now in the excited state $|X\rangle_D$ and is therefore poised to become an emitter that could emit to a new absorber F, or back to the original emitter C, which, now in its ground state, $|G\rangle_C$ serves as a potential absorber. However, there is a gap between D's absorption event A_1 and D's subsequent emission event E_2, because these are distinct events; D plays a different role in each (see Figure 8.5). This reveals that spacetime is not only discrete, but is also a discontinuous structure, in the sense that it consists of independent and distinct photon emission-and-absorption "links." In this respect, the RTI picture differs from the causal set picture in that the chains are not continuous. Figure 8.5 shows several transactional links, and the gaps between them, where the latter involve the continued existence of the participating atoms in the quantum substratum.

8.2.4 Inertial Frames

We now come to an interesting point. Figure 8.5 shows what appears to be a spacetime diagram; however, since the atoms C, D, and F are never "in spacetime,"

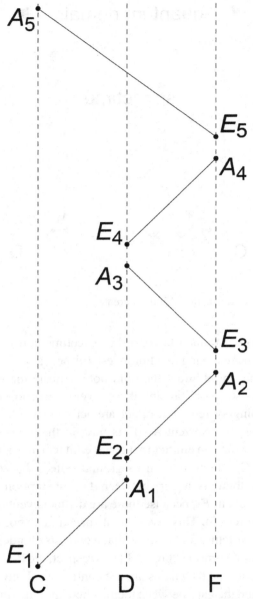

Figure 8.5 Sequential emissions and absorptions.

their "temporal axes" are not really properties of spacetime. Instead, they are *internal* references only. We can think of these as reflecting the counting of internal clocks, that is, strictly local periodic processes that are sequential in nature, such that they yield a locally increasing index. This locally increasing index corresponds to the time coordinate of the atom's rest frame – its proper time. But what is seemingly paradoxical is that this "time coordinate" is not really a property of the spacetime construct! It is just an *internal* reference that is used to index spacetime events from a particular reference frame – the rest frame of the quantum system (such as electron or atom) doing the indexing. And this is why it is not an invariant construct. Thus, we arrive at the following picture: inertial frames are *internal reference structures* of entities in the quantum substratum, *not* aspects of spacetime itself. *Inertia comes from the quantum substratum!* This idea is reinforced by the fact that real photons, which are part of the spacetime construct, do not possess inertial frame.

Spacetime itself is constructed only of invariant quantities: events themselves, and the spacetime intervals or links established by real photons. The irony here is that even though we call this manifold "spacetime," it is not constructed of "space" and "time." These are just frame-dependent parameters used as labels for the connected *events* that actually comprise spacetime. And what connects these events is momentum and energy – really, four-momentum, as contained in the transferred photon. It is in this sense that four-momentum generates spacetime displacements (but only relative to a given inertial reference frame). Energy transfer corresponds to temporal displacement, while three-momentum corresponds to spatial displacement. But "temporal displacement" and "spatial displacement" are *not* themselves aspects of the "spacetime" construct. Both are merely non-invariant descriptions relative to a given inertial frame. The inertial rest frame and attendant "proper time" is defined by the quantum system's rest mass – the invariant quantity and the source of the system's inertia and attendant rest frame.[25]

We can get an idea of what might constitute a physical internal clock corresponding to rest mass by reference to the de Broglie frequency,

$$\omega_{DeB} = \frac{m_e c^2}{\hbar}. \tag{8.1}$$

David Hestenes has constructed a useful model of the electron by incorporating "Zitterbewegung," that is, the fundamental oscillatory motion that is the source of electron spin (e.g., Hestenes, 2010). In this model, the origin of the electron's rest mass is the energy associated with its motion in a light-like helix (see Figure 8.6).

This helical motion defines a time-like "world tube" corresponding to an effective subluminal rectilinear momentum. The model provides correct correspondence with the Dirac theory of the electron. In particular, it obtains a

---

[25] Sir Arthur Eddington makes this point – that spacetime parameters comprise a reference system, not an invariant ontological entity – quite powerfully in his book *The Mathematical Theory of Relativity* (1960, chapter 1, §1.1).

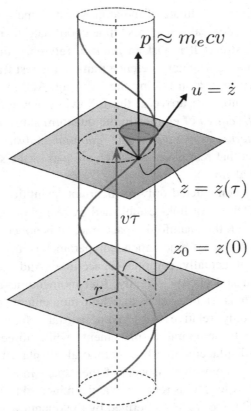

Figure 8.6 The "Zitter" model of the electron by D. Hestenes (2010).
Used with permission.

fundamental angular frequency that modifies the de Broglie frequency with a
factor of 1/2, that is,[26]

$$\omega_e = \frac{2m_e c^2}{\hbar}. \qquad (8.2)$$

The corresponding periodic "clock" function can then be written as

$$\psi(\tau) = e^{i\omega_e \tau} \qquad (8.3)$$

where τ is the electron's proper time. However, it's important to note that this
oscillatory "motion" is *not* spacetime motion. The electron's half-integral spin
involves a nontrivial topology, and we cannot pretend that what is "revolving" is
a classical point particle. Rather, the rotational motion is that of a spinor – a specific

---

[26] Schrödinger originally obtained this frequency by analyzing the time dependence of the velocity operator for
the Dirac equation.

quantum entity that undergoes a sign change upon a rotation of 2π. This means that a "double rotation" of 4π is required for return of the electron to its initial state.[27] In any case, the fundamental frequency in (8.2) provides a periodic process that serves as the "internal clock" that provides the proper time reference for the electron, that is, that defines the temporal axis of its rest frame, and which therefore serves to define the concept of an inertial frame. The concept of an inertial frame is physically distinguished from accelerating frames via the fact that the fundamental frequency (internal clock) is obtained from the Dirac equation for a free (non-interacting) electron.

We digress here briefly to note that the foregoing resolves a long-standing puzzle regarding whether Newton's concept of "absolute space" is needed to define non-inertial motion such as that of the rotating water in Newton's famous spinning bucket experiment. We now see that both forms of motion – inertial and non-inertial – are defined in terms of the quantum substratum, not "absolute" space, time, or a substantive "spacetime continuum." Specifically, inertial motion is defined by reference to non-interacting fermionic systems (or bound systems composed of fermionic fields), while non-inertial motion is defined by reference to interacting fermionic systems. *It is the quantum substratum that constitutes the absolute reference for types of motion.* We return to this issue in Section 8.4.

Similar considerations apply to any quantum system with nonvanishing rest mass, such as larger fermions, atoms, and molecules. It is rest mass that generates the internal clock that serves to define the proper time and thus the temporal axis for the system. Metaphorically speaking, we can think of rest mass as the "sand in the hourglass." Rest mass originates in the nontrivial topology of fermions – quantum systems with half-integral spin. We can think of fermions as forms of "trapped light," since the fundamental structure is that of an electromagnetic field confined to a pre-spatiotemporal topological vortex. Recalling Chapters 4–7, this quantum substratum is a form of physical possibility, so electrons and other quanta with rest mass retain the ability to enter into superpositions. However, they can be "collapsed" into determinate states through participation in transactions. This issue was quantitatively addressed in Section 6.5.

Let us return now to photons, having zero rest mass. Of course, the photon has no rest frame and is not an inertial object. From the vantage point of a photon, no time elapses between its emission and its absorption, so the photon's "clock" is static. Neither is there any spatial separation, from the vantage point of a photon,

---

[27] In addition, Hestenes (2019) notes that the Dirac wave function really describes an ensemble of helices, not a single helix, though he proposes that the electron itself corresponds to one of the helices. In the present interpretation, the electron corresponds to the entire ensemble and should not be thought of as a point particle. This is because the idea of a position continuum is an idealization, and localization is only a phenomenon arising from the tiny (but finite) size of micro-absorbers such as atoms and molecules.

between its emission event and its absorption event. For the photon, there is no distinction between the "time axis" and the "spatial axes" – they are merged as the photon's null internal. The distinction between time and space appears *only* by reference to an object with rest mass that defines an inertial frame. And "rest mass" is, in effect, electromagnetic energy confined to a topological vortex in the quantum substratum.

We noted above that links are established via photon transfers; the emission event is the parent of the absorption event. Considering again Figure 8.5, what about the time-like gap between events A_1 and E_2, in which atom D stands ready to emit after having absorbed? In a chronological sense, A_1 is the parent of E_2, but they are not connected by a transferred photon "link." We can describe this situation with the concept of an *implicit link* (IL). An IL is not part of the spacetime construct or causal set, but it still contains temporal information, including an arrow of time, by reference to the inertial frame defined by the rest mass "clock" of the quantum system in question. Again, in this picture, inertial frames are *not* aspects of the spacetime manifold; they are internal, quantum-level references. Note that this reflects the vital, physical role played by the quantum substratum as the generator of spacetime, both in the active sense (via transactions that create spacetime links) and in the passive sense (by defining inertial frames).

8.2.5 Spacetime as an "Influence Network"

Another proposal for a causal set structure has been offered by Kevin Knuth and collaborators (e.g., Knuth and Bahreyni, 2012). Knuth et al. champion a nonsubstantival, relational view of spacetime, and in that respect their approach has much in common with the RTI picture. They note the prevalence of the usual notion of spacetime as a fundamental, physical container, but then go on to say:

However, more recently, the idea that space-time is neither physical nor fundamental has been growing [Seiberg, N. (2007)]. The idea is that space and time may emerge from more fundamental relations or phenomena.... In addition to the older ideas, such as space as a container or space as a substance, which have mostly dominated our perspectives of space, is the view that space represents a relation between objects.[28]

These authors call their structure a "poset," for "partially ordered set." They assume, like the present author, that the only real elements of spacetime are events and their influences, which map to emission/absorption events and photon transfers in RTI, respectively. Using only the assumption that an observer can be identified with each chain (totally ordered set of events), together with a consistency condition between

---

[28] Knuth and Bahreyni (2012, p. 1).

observers and the "radar formulas" for time-like and spacelike displacements between events, they obtain the Minkowski metric. While this formulation is thus far restricted to 1+1 spacetime, it may be possible to extend it to 3+1.

The RTI ontology differs in some respects from Knuth's poset picture (KPP). As noted in the previous section, in RTI, objects with rest mass possess internal period "clocks," which serve to establish a proper time independently of specific events. While influences are a primitive concept in the KPP ontology, RTI physically specifies the nature of the influences in terms of photon transfers. In addition, in RTI influences are always mutual, since the transferred photon affects both the emitter and the absorber. This contrasts with KPP, in which an observer either influences *or* is influenced by another observer. While KPP assumes that determinate (classically describable) structures are revealed epistemically, from coarse-graining (using a scale that shows less detail), in RTI these emerge at an ontological level, from the non-unitary transactional process that transforms possibilities into actualities. Nevertheless, the approaches have much in common in that spacetime is a secondary, emergent construct that is fundamentally based on relations and interactions between quantum systems. KPP has great promise in that it manages to extract a great deal of information concerning the spacetime structure, including the Minkowski metric, merely by demanding consistency among the chains concerning the sharing of influences and attendant information about events. A step has also been taken toward accommodating general relativity in KPP, by quantifying the effects of mass on the poset structure. The result is an equation of geodesic form (Walsh and Knuth, 2015).

As noted at the beginning of this chapter, the idea that the spacetime manifold is emergent from the quantum level – as opposed to being an omnipresent "container" for all that is physically real – may seem radical, but arguably it is needed for full ontological consistency of the correspondence between the "fuzzy" quantum level, which seems to violate certain strictures of relativity, and the level of determinate spacetime events that unquestionably obeys relativity. We can gain insight into this matter (no pun intended) by considering the relationship of rest mass, as a source of the gravitational field, to the field itself. Note that the Einstein gravitational field equations are directly analogous to the Maxwell electromagnetic field equations, in the sense that the field is determined by its sources:

Maxwell equations (in covariant form):

$$\partial_\mu F^{\mu\nu} = \mu_0 J^\nu$$

Einstein equations (omitting cosmological constant Λ):

$$G_{\mu\nu} = \kappa T_{\mu\nu}.$$

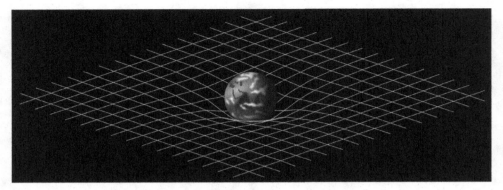

Figure 8.7 A matter source warps spacetime (here represented as a two-dimensional surface) from outside the spacetime manifold.
Source: https://commons.wikimedia.org/wiki/File:Spacetime_lattice_analogy.svg.

In both of these equations, the field generated by its sources is described on the left-hand side, while the sources are represented on the right-hand side. It should be noted that the electromagnetic field sources (charges, J^μ) are not *in* the field. They are sources of the field, and as such, they are not contained within it: field lines terminate at the source charges. Analogously, the proper understanding with regard to gravitation is that the source of the gravitational field – matter – is not contained within the field, which is spacetime (more precisely, the geometric structure of spacetime). The gravitational field (i.e., spacetime and its structure) terminates at its sources, which are material systems. This again tells us that matter is not contained *within* spacetime, any more than charges are contained within the electromagnetic field they generate.

Interestingly, illustrations of the warping of spacetime by matter by projecting the three spatial dimensions down to two depict matter sources outside the spacetime "fabric" (Figure 8.7). Despite the usual uncritical assumption that "everything physical is contained within spacetime," these representations illustrate (however unintentionally) the correct understanding: that matter sources give rise to the structure of spacetime from beyond it. More recently, with three-dimensional animation programs, there have been attempts to depict matter sources curving spacetime in a three-dimensional depiction of the spacetime manifold while being contained within that same spacetime. Yet even in those depictions, which try to place the matter sources "inside the spacetime container," the field structure terminates at the sources, implying that those sources are still outside the spacetime construct.

Einstein himself was uneasy about the relationship between the left-hand side of his equation (expressing the geometric structure of spacetime) and the right-hand side, containing the material energy–momentum tensor $T_{\mu\nu}$, a nongeometric object. As Paul Wesson and James Overduin put it:

The geometrical object $G_{\mu\nu}$ is known as the Einstein tensor, and comprises the left-hand side of the field equations.... However, it is not so widely known that Einstein wished to follow the same procedure for the other side of his field equations. That is, he wished to replace the common properties of matter, such as the density ρ and pressure p, by geometrical expressions. He termed the former "base wood" and the latter "fine marble." (Wesson and Overduin, 2019, p. 6)

According to the current proposal, the reason Einstein's goal of replacing matter by geometry was not achieved is because matter is the source of the spacetime structure, the latter being naturally described by its geometry. In contrast, that which *creates* the spacetime manifold is of a fundamentally different nature, just as electrical charges (electromagnetic field sources) are of a fundamentally different physical nature from the field to which they give rise, and they are neither part of it nor contained within it. In fact, the material sources of the gravitational field (i.e., spacetime) are quantum systems, and as such are not contained within spacetime: quantum systems are physical possibilities, while spacetime is a structured manifold of actualities. Nevertheless, the fact that matter is not part of spacetime does not mean that it is deserving of the apparent contempt ("base wood") in which Einstein held it. On the contrary, arguably it is because of sophisticated and subtle topological and symmetry principles that matter can serve as the source of the spacetime manifold. For example, rest mass arises through fermionic spin, a topological property as noted in the previous section; and charge, which couples with the electromagnetic field and thereby gives rise to photon transfer and transactions resulting in spacetime events, can be understood in terms of a U(1) gauge symmetry. None of these is a property of spacetime, but rather they are properties of the physical possibilities – quantum systems – that are the sources of spacetime.

8.2.6 A Common Worry and Why It's Not a Problem

In this section, I address a question that pops up from time to time as a possible objection to the transactional picture. The scenario involves a very distant star that engages in a transaction with a person's eye, so that they see the star as it existed billions of years ago. But suppose the star has long since ceased to exist, and that it sent out that photon long before this observer was born. How did the star "know" that the observer would be in the right place at the right time to engage in this transaction?

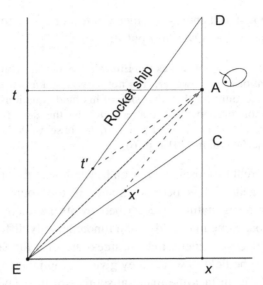

Figure 8.8 A star emits to a person who thinks that the star emitted and died before he was born.

Actually, the star didn't need to know, because it didn't send out the photon long before the person was born. There are two main issues overlooked in the construction of this little paradox:

(1) In view of relativity, distances and time lapses are only relative, as is the time order of spacelike-separated events.
(2) No transaction can be set up without the availability of an absorber, in the present, so that any photon transfer establishes the emission event *in the past*.

First, consider point (1), with reference to Figure 8.8. Assume the star and the person (call him Bob) are in the same inertial frame. Event F is the star's demise. The story in which the star has ceased to exist before Bob comes along only holds relative to certain inertial frames. In fact, there is no invariant distance between the star and Bob, nor is there any invariant time of travel for the photon to get from the star to Bob. There is also no invariant time order of the events involving Bob's birth (denoted by B) and the star's emission event, since these are spacelike-separated. Consider a rocket ship traveling very fast from the star toward Bob. Its spatial axis (compressed to one dimension) is the slanted line intersecting point C. This means that point C is simultaneous with the star's emission according to the rocket's perspective. From Bob's perspective, he was born after the star emitted, but from the rocket's perspective, Bob was born before the star emitted.

According to the rocket, the spatial distance from the star to Bob is x', and the time it takes for the photon to reach Bob is t'. These are much smaller than the values x and t assigned by Bob. Thus, from the rocket's perspective Bob is much closer to the star, and the time of the photon's travel is correspondingly reduced. In addition, from the standpoint of the rocket, the star dies only after Bob becomes available as an absorber (well after C, as can be seen by drawing a line parallel to the x' axis from F to Bob's timeline). The lesson here is that the ability of a source to engage with absorbers is not restricted by sequences of events or spatiotemporal displacements relative to any particular inertial frame, since those are not absolute conditions. In this case, we see that according to the rocket ship, there is nothing perplexing about the star engaging in a transaction with Bob.

Regarding point (2), the advent of incipient transactions is governed by the absorbers in the present, and the actualized transaction acts to extrude the new spacetime interval from the present into the past, as a new element of the "spacetime fabric." In this sense, all transactions have a form of built-in retrocausation, but it is limited to the establishment of new spacetime events. It is not an influence contained within spacetime that affects or alters already-actualized events. Emitters and absorbers negotiate in the present (which we can identify with the quantum substratum) via OW and CW, and it's only at the final stage of an actualized transaction that a "past event" is established corresponding to the actualized emission event. So, again, generation of OW and CW, which act in the quantum substratum as a form of *res potentia*, must be carefully distinguished from the actualized real photon that is a spacetime entity. The latter is a form of *res extensa*, as the connection between actualized emission and absorption events. This real photon is represented by a projection operator (outer product of the OW and CW components corresponding to the actualized transaction). Again, the actualized *events* that make up spacetime are *activities* of emitters and absorbers; the latter never become part of the spacetime manifold, remaining in the quantum substratum. In this sense, they are "eternally present."

8.3 Transactions Break Time Symmetry and Lead to an Arrow of Time

In this section, we look more closely at the issue of the directedness of the emergent spacetime manifold and see how the breaking of temporal symmetry through the transactional process or NU-interaction provides a natural basis for the "arrow of time" that is so apparent in macroscopic thermodynamics.

The Second Law of Thermodynamics unambiguously describes irreversibility and an "arrow of time." Specifically, it states that the entropy S of a closed system can never decrease: $(ds/dt) \geq 0$. Roughly speaking, entropy is a measure of "disorder," that is, how far a system is from a state in which it contains information in the form of an ordered configuration. At a maximal value of entropy, we have an equilibrium state – a state containing no specific information. An example is a box of gas in which different kinds of gas molecules are completely mixed together so that there is no available information about the whereabouts of the different gases.[29] We routinely see this law corroborated at the macroscopic level of everyday experience, where we find S increasing with increasing time in closed systems – that is, we see closed systems evolving toward equilibrium, but never the opposite. However, since it is commonly assumed that the laws of physics are time-reversible, the physical origin of this time-asymmetrical law has remained a mystery (despite numerous less-than-successful attempts to account for it).

Of course, we have already seen that in the RTI picture, quantum measurements result in an arrow of time, since they establish new spacetime events in a future-directed manner. In particular, in an actualized transaction, the emission event always precedes the absorption event chronologically. The temporal asymmetry can also be found in the representation of creation and annihilation in standard quantum field theory, which is formally recovered from the direct action theory as discussed in Chapter 5. Emission of the real photon corresponds to creating a Fock state by operating on the vacuum state with a creation operator, that is, $|k\rangle = \widehat{a}_k^\dagger |0\rangle$, while absorption of the real photon corresponds to operating on a Fock state with an annihilation operator, that is, $\hat{a}_k |k\rangle = |0\rangle$.

The above is an inherently asymmetrically process, since something must be created *before* it can be destroyed. That is, if one tries to annihilate something that doesn't exist, one gets no state at all – not even the vacuum state: $\hat{a}_k |0\rangle = 0$. This temporal directionality comes from the broken temporal symmetry inherent in the quantum boundary condition discussed in Chapter 5.[30] That is, in order to have any real photon, the symmetry of field propagation must be broken by absorber response (or, at the relativistic level, the NU-interaction). The real photon is a time-asymmetric entity; it proceeds *from* an emitter *to* an absorber in a temporally directed manner. But its creation also involves an irreducibly stochastic process, the non-unitary interaction. We consider this aspect in more detail below.

8.3.1 The Origin of the "Initial Probability Assumption"

Ludwig Boltzmann famously attempted to derive the Second Law in the context of statistical mechanics. The result was known as the "H-theorem," but it crucially

---

[29] More precisely, however, entropy S is defined in terms of a quantity of heat Q relative to a particular temperature: the change in entropy is $dS = dQ/T$.

[30] At this point, the reader might object that time-asymmetry is being "smuggled in" through the boundary condition of the direct-action theory that yields the Feynman propagator. But as discussed in Chapter 5, note 13, this is not an ad hoc future-directed condition, since the theory has two semi-groups corresponding to the Feynman and Dyson propagators, each of which corresponds to the same empirical phenomena (where the sign of the energy and "increasing time" ends up only a convention). In contrast, standard quantum field theory does in fact impose an ad hoc time/energy asymmetry, through the choice of the invariant phase space element with positive energy, $d^3p/(2\pi)^3(2E)$.

depended on what at that time was an ad hoc irreversibility assumption, which he termed the "assumption of molecular chaos." The German term was the *Stosszahlansatz*. This consisted of assuming that state transitions among atoms and molecules were fundamentally stochastic and dependent only on their most recent interaction – a Markov assumption. This statistical assumption appears in various forms in all attempts to derive the Second Law from what are otherwise assumed to be deterministic and reversible physical laws. It crucially appears in the use of "master equations," which describe state transitions by indeterministic, stochastic processes. Pauli's ad hoc "random phase assumption" (Pauli, 1928) was the quantum-equivalent of the *Stosszahlansatz*.

Sklar has noted that "[t]he status and explanation of the initial probability assumption remains the central puzzle of non-equilibrium statistical mechanics" (Sklar, 2015). In this section, we'll look at the details of the origin of what Sklar terms the "initial probability assumption" that yields the Second Law. Crucially, under RTI, this ceases to be an ad hoc assumption. The probabilistic description has the same fundamental source as the directionality of the growing spacetime construct, namely, the non-unitary interaction (NU-interaction, or relativistic equivalent of "absorber response") discussed in previous chapters.

Traditionally, quantum theory has involved the use of a so-called statistical operator or "density operator," $\hat{\rho}$, where

$$\hat{\rho} = \sum_i P_i |\Psi_i\rangle\langle\Psi_i|. \tag{8.4}$$

In (8.4), $|\Psi_i\rangle$ are pure states, and P_i is the epistemic probability that the system under study is in state $|\Psi_i\rangle$.[31] In general, $\{|\Psi_i\rangle\}$ is not a basis, since these states need not be mutually orthogonal. Under the usual Schrödinger picture, the time-dependent Schrödinger equation and its adjoint yield a unitary (deterministic) time dependence of the density operator $\hat{\rho}$:

$$\frac{\partial\hat{\rho}}{\partial t} = \frac{-i}{h}[H, \hat{\rho}]. \tag{8.5}$$

However, it's important to note that $\hat{\rho}$ is *not* an observable. It is based on the Schrödinger picture of quantum mechanics, which attributes temporal evolution to the quantum state. Its time-dependence is opposite to that of an observable O in the Heisenberg picture, which obeys the equation:

---

[31] Though it's standard practice to assign pure states to systems of interest, under the usual assumption that quantum theory "really" only has unitary dynamics, this involves ignoring the important distinction between a proper and improper mixture. Under the unitary-only assumption (rejected in TI), pure states are never applicable to a given quantum system except by appeal to an ad hoc partition of the Hilbert space such that the system of interest is assumed to be in a product state with other "environmental" degrees of freedom.

$$\frac{\partial \hat{O}}{\partial t} = \frac{i}{h} \left[H, \hat{O} \right].$$ (8.6)

As discussed in Chapter 6, Section 6.5, the deterministic, unitary evolution described in (8.5) only applies to systems with nonvanishing rest mass – for example, emitters and absorbers – and only between transactions, the latter constituting non-unitary measurement interactions.[32] As we have seen in previous chapters, the non-unitary transactional NU-interaction irreversibly projects the system into a proper mixed state, diagonal in the basis corresponding to the observable measured, with outcomes weighted by the Born Rule. This constitutes a deviation from the evolution in (8.5). Thus, for finite rest-mass systems, the quantum evolution is really a form of a "piecewise-deterministic Markov process," since there is deterministic (unitary) evolution interrupted by non-unitary, stochastic behavior based on the latest state of the system, which constitutes a Markov process.

The Markovian behavior arises from the transactional NU-interaction constituting the process of measurement, in which one or more photons is irreversibly transferred from emitters to absorbers that are otherwise described by (8.5). The non-unitary interactions of RTI can thus be readily identified with the stochastic thermal processes assumed by traditional thermodynamics, whereas such processes cannot be truly stochastic under the usual assumption of universally deterministic evolution. Thus, the picture here is that irreversible measurement-type interactions (transactions) frequently and repeatedly interrupt the Schrödinger deterministic evolution of entities such as gas molecules, naturally yielding the genuinely stochastic behavior that is required for justification of the "initial probability assumption."

For example, in Section 6.5, we discussed an example of an emitting atom that, together with a set of responding absorbers, creates a set of incipient photon transactions corresponding to various values of directional momenta **k**. Under RTI, these incipient transactions are weighted by the Born Rule probabilities, which are objective in nature. That is, there is no deterministic law that dictates which value of **k** will be actualized; the process of reduction or "collapse" is a form of spontaneous symmetry breaking. This is the fundamental origin of the Markovian behavior, which is irreducibly stochastic. Markov behavior contradicts the usual assumption of universally deterministic evolution, leading to the apparent inconsistency of the "initial probability assumption" critiqued by Sklar but which is crucial for deriving the Second Law. The fundamentally stochastic behavior is unproblematic under TI, since the irreversibility of the transformation from a pure

---

[32] Unitary evolution also applies to virtual (off-shell) photons, which effectively possess rest mass.

state into a proper mixed state (von Neumann's "Process 1") and its stochastic nature is a basic feature of TI. Thus, under TI, stochastic behavior at the micro-level ceases to be an unjustified ad hoc assumption – which it always must be in the context of any fully deterministic (unitary) theory. We therefore see that Boltzmann's *Stosszahlansatz* gains a specific physical explanation in TI.

8.3.2 "Master Equations" Justified in TI

Once we understand that quantum systems really undergo stochastic processes that interrupt the deterministic evolution of (8.5) and clearly define a measurement basis, the origin of the "master equations" that feature prominently in derivations of increasing entropy becomes clear and physically well grounded. A master equation is a differential equation relating the change in the occupancy probabilities of relevant physical states (such as states of well-defined energy) to the transition probabilities to and from those states. It uses the principle of "detailed balance" – the condition that the occupancy probabilities remain static at equilibrium. A master equation presupposes fundamentally stochastic, Markov processes, and a specific, stable measurement basis. In other words, it presupposes a classical, Boolean probability space that is not available under standard, unitary-only quantum theory, but which is provided in TI's account of the measurement process.

An example of a master equation is:

$$\frac{dP_i}{dt} = \sum_{j \neq i} R_{ij} P_j - R_{ji} P_i \tag{8.7}$$

where P_i is the probability that the system is in state $|i\rangle$, and R_{ij} is the transition probability from state j to state i. Equations such as (8.7) make intuitive sense because the probability of occupancy of a state i increases with the transition probability from other states j into that state (described by $R_{ij}P_j$) and decreases with the transition probability away from that state into other states (described by $R_{ji}P_i$). A more formal derivation of expressions like (8.7), which reveals the Markov assumption, is given in Toral (2015).

As illustration, consider a simple two-state system in which the transition probabilities R_{ij} between states 1 and 2 are both 1/2. The solutions of the differential equation in this case will be

$$P_1(t) = \frac{1}{2} + \frac{P_1(0) - P_2(0)}{2} e^{-t} \tag{8.8a}$$

$$P_2(t) = \frac{1}{2} + \frac{P_2(0) - P_1(0)}{2} e^{-t}. \tag{8.8b}$$

These solutions for the occupancy probabilities of each state demonstrate an approach to equilibrium. For large times, the second term with dependence on the initial state vanishes, and we are left with the constant term that depends only on the transition probabilities. Typical "derivations" of the Second Law help themselves to such master equations to "demonstrate" approach to equilibrium. But under the usual assumption that the underlying physics is fully unitary (and therefore reversible), application of the probabilistic description of the master equation can only come from so-called coarse graining over deterministic Liouville phase space trajectories. Coarse graining amounts to simply ignoring the putative deterministic behavior of the unitary-only account, thus introducing only an epistemic (observer-dependent) uncertainty that does not describe the physical behavior of the system itself. This is inadequate, since the prediction of approach to equilibrium depends on the applicability of master equations that describe physical systems as behaving in a genuinely stochastic, Markovian manner (the content of Boltzmann's *Stosszahlansatz*). We get this only from a theory in which the microscopic behavior is genuinely Markovian, such as the transactional account. Thus, if nature in fact behaves in accordance with TI, then Boltzmann was right in invoking the *Stosszahlansatz*; that is, it ceases to be an element of logical circularity (i.e., assuming the irreversibility we want to derive) and instead describes a real behavior of physical systems. Thus, determinism and reversibility are broken in the transactional picture in exactly the right way to gain a noncircular account of the approach to equilibrium and the Second Law of Thermodynamics.

8.4 Spacetime Relationalism

I noted in Chapter 4 and above that RTI assumes a *relationalist* view of spacetime. In this section, I examine relationalism in more detail as a position in opposition to spacetime *substantivalism*, following aspects of the formulation in Friedman (1986).

The spacetime substantivalist views spacetime as a substantive manifold M of points $\{a, b, c, \ldots\}$, each indexed by the temporal coordinate t and spatial coordinates (x, y, z), where all the indices are real numbers ranging from minus infinity to plus infinity. The manifold itself is considered to have structure in the form of symmetries and a metric. In particular, according to relativity theory, the square of the spacetime interval ds is a real-valued function $I(a, b)$ defined on M. The key point is that according to substantivalism, not all of the spacetime points correspond to physical events; rather, only those points belonging to some subset P of M are occupied by concrete physical events.

In contrast, the relationalist thinks that there is no substantial spacetime manifold M but that there are *only* concrete events whose collective features contain all the necessary qualities to account for the observed symmetries and phenomena conventionally associated with spacetime itself. While this work does not attempt to present a case for relationalism (which has been ably provided by numerous authors[33]), it seeks to place RTI in the context of the discussion concerning the competition between these two views. As Friedman (1986) has noted, relationalism has no significant challenges in accounting for the symmetry aspects of spacetime; indeed, it has advantages over the substantivalist view in that regard. However, researchers have generally been unable to find in the relational approach the "absolute background" that is traditionally seen as necessary for formulating laws of motion. This shortcoming is remedied in RTI, since as we have seen in Section 8.2.4, it is the quantum substratum itself that defines inertial frames and therefore functions as the absolute background required.

Friedman distinguishes two main facets of relationalism in the literature: what he terms *ontological* and *ideological*. They are defined as follows:

"Ontological": Spacetime is no more than the set of existing events P.
"Ideological": Existing spacetime events meet certain physical requirements (such as "causality").

These two approaches are not different versions of relationalism but rather aspects of it that are primarily under debate. For example, as Friedman notes, "ontological" considerations were primarily at issue in the Newton–Leibniz debate, which concerned Newton's postulation of an "absolute space" and "absolute time" held by Leibniz to be without legitimate physical content; while "ideological" considerations have been at issue in the more recent discussion revolving around Reichenbach's and Grunbaum's contributions.[34] RTI's relationalism could be said to address primarily the "ideological" aspect in that it defines eligibility for membership in P (the set of concrete events) in terms of a specific physical process: the transaction.

As noted previously in this work, most practicing physicists believe very strongly in spacetime as a substance, that is, as an entity that exists in its own right as a dynamic "container" which supports events and influences their interactions. Yet spacetime itself is not observable. There is no actual empirical evidence of the independent existence of a spacetime substance as something distinct from events. Rather, the existence of spacetime is *inferred* from observable phenomena based (in large part) on the metaphysical view that events require a "container," that is, the view that it is not enough to say just that events themselves (and their collective

---

[33] For example, Sklar (1974), Barbour (1982), Earman (1986), and Brown (2005).
[34] Cf. Reichenbach (1958) and Grunbaum (1973).

structure) exist. The general theory of relativity is often invoked in support of a substantivalist view, since it relates the metric characterizing sets of events to the mass–energy of the fields instantiating the events. But the same can readily be understood in terms of a relational (antisubstantivalist) view of spacetime (cf. Brown, 2002, p. 156). The basic point is that spacetime is an extra-empirical notion, in the same sense that causality is an extra-empirical notion: neither is actually observed nor directly referred to by theoretical entities.

Now, the reader might protest: "Surely, spacetime *is* referred to because many entities, such as fields, contain spacetime arguments: for example, $\Psi(x, t)$." However, there are (at least) two reasons why the use of such arguments does not constitute a reference to a spacetime substance: (1) the arguments (x, t) are not invariant; that is, they are dependent on the state of motion of the observer; and (2) they are defined only relative to distances (intervals) between events, or to an arbitrary coordinate system, not in an absolute sense. Both (1) and (2) imply that (x, t) refers to *a relationship between events* (and/or observers) rather than to something external to those events. While practitioners of quantum field theory often characterize the theory in terms of the association of a field with "all points in space," that formulation gratuitously adds to the theory the uncritical presumption that space is a preexisting substance.

Finally, we should note that Einstein himself denied the idea of a spacetime container: "There is no such thing as empty space, i.e., a space without a field. Spacetime does not exist on its own, but only as a structural quality of the field" (Einstein, 1952). Since matter is the source of the gravitational field, Einstein's expressed ontology is completely harmonious with the RTI picture, in which events and their connective structure emerge from the interactions among rest-mass systems in the quantum substratum. That is, the "field" is the structured set of events established through actualized transactions, not a substantive continuum of unoccupied points.

8.5 RTI versus Radical Relationalism

RTI, while eliminativist about time as a substance, provides a useful compromise between the radical relational (RR) view of Carlo Rovelli (1996) and the substantival view of time (i.e., that time is a real entity). Rovelli's picture, subject to the usual unitary-only assumption, follows the quantum state through its progressive interactions with measuring devices, so that the linearity of the state carries over to the macroscopic scale. Thus, in Callender's words:

Consider the famous case of Schrödinger's cat. The cat is suspended between life and death, its fate hinging on the state of a quantum particle. In the usual way of thinking, the cat becomes one or the other after a measurement or some equivalent process takes place. Rovelli, though, would argue that the status of the cat is never resolved. The poor thing may be dead with respect to itself, alive relative to a human in the room, dead relative to a second human outside the room, and so on.... It is one thing to make the timing of the cat's death dependent on the observer, as special relativity does. It is rather more surprising to make whether it even happens relative, as Rovelli suggests, following the spirit of relativity as far as it will go. (2010, p. 64)

This is indeed the logical conclusion of applying a relational view to quantum mechanics if one does not take absorption and non-unitarity (only available in the direct-action picture) into account. RTI differs from both RR and the "usual way of thinking" regarding nonrelativistic quantum mechanics, in that it provides for a definite empirical result to occur; that is, an event is actualized (via a transaction) which renders the cat definitely alive or definitely dead. Under TI, the measurement process is clearly defined as explained in Chapters 3–5: it is precipitated by the NU-interaction, which is governed by a clearly defined probability. As discussed in Section 5.3.3, since there are so many absorbers in a macroscopic situation, the measurement is completed long before any macroscopic object could be placed into a linear superposition (in particular, a superposition involving different localizations). Note that this situation is still perfectly consistent with relativity in the sense that the spacetime coordinates given to the actualized event are relative to an observer. Two observers in different inertial frames will disagree on the coordinates of the event, but will agree on the spacetime interval between that event and another event (and on what those events are). In RTI, as opposed to RR, specific empirical events do exist; it is only their individual spacetime "location" which is relative, which reflects the fact that spacetime does not exist as an independent entity.[35] In this regard it is useful to consider Callender's apt analogy between time and money[36] as secondary, derivative notions:

In Einstein's thought experiments, observers establish the timing of events by comparing clocks using light signals. We might describe the variation in the location of a satellite around earth in terms of the ticks of the clock in my kitchen, or vice versa. What we are doing is describing the correlations between two physical objects, minus any global time as intermediary. Instead of describing my hair color as changing with time, we can correlate it with the satellite's orbit. Instead of saying a baseball accelerates at 10 m/s, we can describe it in terms of the change of a glacier. And so on. Time becomes redundant. Change can be described without it.... This vast network of correlations is neatly organized, so that we can define something called "time" and relate everything to it, relieving ourselves of the burden of keeping track of all those direct relations.... But this convenient fact should not trick us

---

[35] While I focus here on the unreality of time, the basic relational view is that the spatial component of spacetime is nonfundamental as well.

[36] Hence the equivalence often cited between the two: they are both equally illusory.

into thinking that time is a fundamental part of the world's furniture. Money, too, makes life much easier than negotiating a barter transaction every time you want to buy coffee. But it is an invented placeholder for the things we value, not something we value in and of itself. Similarly, time allows us to relate physical systems to one another without trying to figure out exactly how a glacier relates to a baseball. But it, too, is a convenient fiction that no more exists fundamentally in the natural world than money does. *(p. 65)*

Interestingly, under RTI, what quantum mechanics in fact does is to "negotiate a barter transaction" every time a quantum is emitted and absorbed. While it is too much trouble for us to keep track of all this bartering (as Callender notes), nature performs this complicated bookkeeping task admirably, which is why the vast network of correlations is so "neatly organized." The events themselves are actual for everyone; it is only their spacetime descriptions that are relative. Therefore, while I admire the spirit of Rovelli's exploration, I think it is not necessary to deny that clearly defined events exist. Relationalism need not deny that specific events exist; it need only deny that some independent substance called "spacetime" exists. Indeed, the core of relationalism is that it is the structure of the collection of events that defines what we think of as "spacetime."

8.6 Ontological versus Epistemological Approaches, and Implications for Free Will

The metaphysical picture proposed here may seem strange or "far-fetched."[37] But there is nothing inconsistent about it and much to recommend it as a viable ontology underlying quantum theory. If one takes the mathematical objects such as state vectors as referring to something ontologically real (as opposed to being just a measure of our knowledge or ignorance), then the entities and processes to which they refer obviously cannot "fit into" spacetime and therefore, if real, must exist in some pre-spacetime realm (the quantum substratum). This, again, was noticed even by Bohr in his previously quoted comment (see Section 2.5) that such processes must "[transcend] the frame of space and time."

Nevertheless, one might consider whether we should resort instead to an epistemic-type interpretation of quantum states, as in the time-symmetric "hidden variables" approach of Price (1996) or the models studied by Spekkens (2007).[38] Such interpretations imply a "block world," that is, that events are already "there" in spacetime and that various types of "hidden variables" (i.e., unknown aspects of

---

[37] Of course, "many worlds interpretations" can certainly be considered at least as "far-fetched," so one should be careful to avoid a double standard here. We should also keep in mind that it was considered far-fetched for Galileo to insist that the earth was in motion when any one of his contemporaries could clearly "observe" that it was not moving. Appearances can be deceiving.

[38] As noted in Chapters 1 and 2, Spekkens' models may, in any case, run afoul of the Pusey et al. theorem (2011).

the "ontic state" of a system) encode additional information about which events, out of an *apparent* choice among possible events, are actually "chosen already."

I believe that such an approach – taking quantum theory as "incomplete" – misses a valuable opportunity to discover what truly novel message might be contained in quantum theory, as discussed in Chapter 2. For one thing, such approaches (e.g., the Bohm model) have ongoing difficulties with the relativistic domain, while RTI is fully compatible with it.[39] Moreover, there would appear to be no room in the "block world" implied by a hidden variables approach for the experience of human agents as having free will concerning what they choose to measure or to create; there is no *genuine* becoming. Of course, this brings us back to the age-old philosophical discussion concerning fatalism versus free will, and I do not pretend to do justice here to this intricate, long-standing debate. However, it is generally accepted that in a strictly predetermined world, there can be no genuine free choice in the sense of an unconstrained selection of one path from a "garden of forking paths," since there are no forking paths. In order to "save the experience" of free will, one has to resort to an argument that one can "freely" choose what one is already destined to choose.[40] If in fact choices are already made and already there in the block world, then there are no real "choices" at all, and our perception that we are really making free choices is an illusion.

Nevertheless, a recent trend seems to be emerging among some interpreters of quantum theory: the idea that physics implies that there is a block world and that the correct interpretive task should therefore be to examine the ramifications of that ontology. As noted above, I believe that this is a mistake based on taking a particular kind of map for the actual territory. If one believes that the block world model is correct, one consequence that follows is that the events that we see unfolding around us as we "move along our worldlines" don't actually "happen" in any particular order, that is, that all events simply exist in the block world and that therefore the direction of events is arbitrary and a matter of perspective. (Note that this view also depends on a primitive assumption that we are "moving through" the block.) A block world adherent might assert, for example, that the directional quality of events is simply a matter of the kind of creatures we are, that is, that

---

[39] Sutherland (2008) proposes a time-symmetric version of Bohm's theory that has formal compatibility with relativity, although its final-time boundary condition must single out a preferred inertial frame. This constitutes another type of epistemic, block world–type interpretation.

[40] This position is known as "compatibilism," the view that determinism is compatible with free will. While much of the free will versus determinism debate concerns moral responsibility and is therefore beyond the scope of this work, the basic compatibilist argument boils down to the idea that free will just means being able to act in accordance with one's wishes in an unfettered manner. Given determinism, one's wishes must, of course, be fully determined by laws and factors beyond one's control, so this definition seems vacuous. This move also fails for the situation in which an experimenter has an apparent choice between two possible measurements but no personal preference of one over the other; one cannot mitigate the force of determinism against free will by appealing to one's wishes or desires in such cases.

some other kind of creature in the block world would see things entirely differently. A fictional example might be Merlin the Magician, a different kind of creature who is facing in the opposite direction and "moving through" the block in the opposite direction.[41] One can even imagine picking up the "block world" and turning it sideways by ninety degrees, that is, interchanging space and time. But this disregards the important metrical distinction between the space and time indices and the fact that there is no quantum mechanical time operator while there is a position (space) operator, if only at the nonrelativistic level.

Along with the block world approach goes the assumption that many of our "intuitions" about the world must be inaccurate. Among these are (1) our experience of only one event at a time, (2) the perception of "nowness," (3) the perception that radiation proceeds *from* an emitter *to* an absorber in a diverging spherical wave, and (4) the sense that we have free will, that is, the capacity to intervene in events and alter their future courses by our choices. However, it should be noticed that at least some of these so-called intuitions, for example, (1) and (3), are in fact well-corroborated empirical observations.

As is evident at this point, I disagree that one should take the block world ontology as the message of physics. Granted, we may need to revise some of our "intuitions," such as the idea (discussed in Chapter 2) that we can perceive everything that exists. However, we need to be careful that in jettisoning what might be called "intuitions," *we are not actually jettisoning the empirical reality that physics is supposed to be explaining.* For surely the world of appearance, as reflected in much of the list of perceptions above (but perhaps not including free will), *is* the empirical realm. Since there is an interpretation of physical theory (the one I propose herein) that can explain, rather than deny, many aspects of empirical reality, surely that is methodologically preferable to taking one kind of map as the actual territory and embracing the consequence that we are radically and collectively mistaken about our thoroughly corroborated empirical observations, such as the direction of radiation flow.

The point is that a less radical option is available: simply admit that there may be subempirical entities that we are not able to perceive at the empirical level, and that those are the objects to which quantum theory refers. In fact there is ample precedent – that is, Boltzmann's atomic theory, so despised by Mach the staunch empiricist, but subsequently vindicated by its fruitfulness – that this is the best option.

In contrast to interpretations that take quantum theory as referring to observers' ignorance concerning already established spacetime events, RTI does not have to

---

[41] It is important to note that Merlin would be in category III of Chapter 7's possibility types; that is, he is no more real than "that possible fat man in the doorway."

sacrifice genuine free will or make do with an impoverished "illusory free will" substitute, because it is fully harmonious with free will. In RTI, no spacetime events exist apart from actualized transactions. So, for example, the fatalist argument rehearsed in Dummett (1964) does not apply. Dummett's challenge concerns statements about the future. He argues that such statements must refer to something in order to be either true or false and that their referents are future events, which must therefore exist. However, in the current proposal, sentences about the future such as Dummett's example, "I will be killed in the next air raid," do not refer to spacetime events; they refer only to possible events. Such statements are genuinely neither true nor false because they refer not to preexisting events but only to possibilities in the quantum substratum – which are *objectively* uncertain.

Note that the above response of RTI to fatalism is not reducible to the claim that the sentence above is neither true nor false at the time it is uttered but will become either true or false at the time to which it refers, in response to which the fatalist can just rephrase the above sentence as "The *statement* about my being killed in the next raid will either become true or false." This is because *both* statements refer only to possible events in the quantum substratum,[42] not to actualized events in spacetime. That is, the sentence "The statement about my being killed in the next air raid will either become true or false" is just as much a statement about an objectively indeterminate future as is the original statement, "I will be killed in the next air raid." Both statements ultimately refer to *both* sets of alternative possible events in the quantum substratum: a subset in which I am killed and a subset in which I am not killed (assuming there actually will be an air raid – which is also objectively uncertain!). Thus RTI can deny fatalism while retaining the meaningfulness of statements about the future in terms of real, but objectively uncertain (i.e., unactualized), events.[43] There is no "fact of the matter" about whether I will be killed tomorrow when I am making statements (or statements

---

[42] Technically, quantum systems and their interactions exist in the quantum substratum, but those are precursors to events, so as a kind of shorthand, we can refer to the quantum substratum as containing "possible events."

[43] Dummett (1964) expresses skepticism that one can deny bivalence (i.e., either truth or falsity; no "middle" option) for future-tense statements (in fact, he characterizes the response necessary to avoid the fatalist argument as a denial that there can be a "genuine" future tense). But that particular exposition presupposes classical notions about spacetime which one should be prepared to reevaluate in the face of quantum theory. Moreover, one can question the implicit premise of passivity contained in the air raid example and other examples used to argue for fatalism. If there is genuine free will, then creatures with free will (such as humans) can actively participate in the "weaving" process that is the creation of spacetime events. So, even if one wants to keep bivalence (truth or falsity) about future events, one can meaningfully talk about such future events as objectively uncertain but as definitely taking place or not "when the time comes": for example, I freely may or may not choose to bring about a particular event; but if I do, it definitely occurs, and if I do not, it definitely does not occur. The fact that it ultimately either occurs or does not occur does not mean that my fate was "sealed" at any time prior to that event's actual occurrence (or non-occurrence).

about statements) about tomorrow's air raid.[44] In the quantum substratum, all statements about the future are meaningful, but are objectively uncertain as to truth value, because that to which they refer is objectively uncertain.

In the view of this author, taking quantum theory as "incomplete" leaves us with a rather impoverished ontology in which humans must be radically and collectively deceived about their ability to choose and to create. The advantage of RTI over a static block world view is that the networks of transactions retain a kind of crystalline beauty sometimes attributed to the block world: transactions certainly express relevant spacetime symmetries. Therefore, one can still have the aesthetically appealing symmetries without sacrificing a thoroughgoing realist approach that provides additional richness to the ontology, rather than (as in the block world picture) *subtracting* ontological content by denying that the state vector fundamentally refers to something that exists in the world.[45]

RTI accounts for the empirical spacetime realm in terms of actualized transactions while providing a straightforward basis for subjective experience and free will in terms of a pre-spacetime realm of dynamic possibilities. The connection with the mental realm is not obligatory; RTI is agnostic concerning a relationship, if any, between those possibilities and mental activity. But if there is an empirically unobservable realm transcending the spacetime realm of appearance, that would seem to be a prime candidate for future research concerning a possible connection between subjective experience and quantum theory. For further details on the viability of robust free will in the context of quantum theory, see Kastner (2016c).

---

[44] A variant on the block world view is an indeterministic block world, but this is subject to fatalism based on the basic block world assertion that there must be a fact of the matter about any statement about the future.

[45] In case one might argue that "adding richness to the ontology" runs afoul of Occam's razor (OR), my response would be that OR applies to the *methodology* of RTI: quantum theory simply refers to an underlying reality which includes advanced states. This is the simplest explanation of the form of the theory, including the Born Rule. Adding richness to the ontology is evidence of fruitfulness of the interpretation (just as the atomic hypothesis was a fruitful one), rather than an unwarranted complication.

9

Epilogue

I conclude with another quote from Richard Feynman, this one from his Nobel Prize Lecture of 1965:

The chance is high that the truth lies in the fashionable direction. But, on the off-chance that it is in another direction – a direction obvious from an unfashionable view of field theory – who will find it? Only someone who has sacrificed himself by teaching himself quantum electrodynamics from a peculiar and unfashionable point of view; one that he may have to invent for himself.[1]

The irony, of course, is that Feynman himself had championed just such a "peculiar and unfashionable point of view" in his absorber theory of radiation with John Wheeler, a theory which he abandoned. Had Feynman not given up on the absorber theory, we would probably be much farther along in surmounting the challenges presented to us by quantum theory. This issue is discussed in more detail below. First, we summarize what has been presented.

This book has discussed the key ideas and recent developments of the relativistic transactional interpretation (RTI) of quantum theory, which is a relativistic generalization of the original TI of John G. Cramer (1986). A previous edition of this book emphasized the possibilist ontology and used the term "possibilist TI" (PTI) for the same proposal.

RTI is based on the so-called direct action or "absorber" theory of fields proposed by John A. Wheeler and Richard Feynman (1945, 1949), and Paul Davies (1970, 1971, 1972). It provides a solution to the measurement problem of quantum mechanics, in that it allows us to define measurement from within the theory itself. This is possible for RTI because field propagation in the direct-action theory is a relational, mutual interaction, unlike the standard approach to field

---

[1] "The development of the space-time view of quantum electrodynamics," Nobel Lecture (Feynman, 1966).

propagation which assumes that field propagation is a unilateral process. The mutual participation of both emitters and absorbers takes both unitary (force-mediating) and non-unitary (radiative, energy transferring) forms. It is the latter that defines "measurement" through the quantum relativistic version of absorber response, which we have termed the "NU-interaction." The specific physical conditions for each of these types of dynamics were discussed in Chapter 5. Since RTI has non-unitarity "built in," it provides a theoretically grounded basis for the non-unitary "measurement transition" of von Neumann, which has been widely assumed to be a mysterious, nonphysical process that is necessarily outside the theory.

Concerning measurement in quantum theory, Nicolas Gisin noted that

the quantum measurement problem [is] a serious physics problem. Serious because without a resolution, quantum theory is not complete, as it does not tell how one should – in principle – perform measurements. (Gisin, 2017)

The foregoing treatment has shown how the transactional picture yields well-defined physical principles that operate in measurement, thereby addressing and resolving the problem Gisin so clearly identifies. While the absorber theory is empirically equivalent to standard quantized field theories at the level of probabilistic predictions, it is actually a different theoretical model of field behavior, in that the elementary field is nonquantized and constitutes a direct connection between field sources. Quantization of the field arises as a secondary feature, due to the mutual, non-unitary process (known as "absorber response" at the nonrelativistic level) that gives rise to real, on-shell photons. Since RTI is based on a different theory of field behavior, it is technically more than just an interpretation of quantum theory. It amounts to a slightly different form of quantum theory from the standard approach, even though it is empirically equivalent at the level of the predicted probabilities. It is the difference in the account of energy transfer via the fields that allows RTI to explain measurement in physical terms, whereas measurement (as a process distinct from unitary interactions) cannot be explained in physical terms within the standard approach to the theory. The inability of the standard theory to account for the nature of measurement, and the occurrence of single outcomes upon measurement, makes the standard theory empirically deficient; this is the only empirical point of departure between the two theories. Thus, RTI is not at all like other "explicit collapse" theories such as the Ghirardi–Rimini–Weber (GRW) theory, since it does not add *ad hoc* nonlinear dynamics to the Schrödinger evolution that, in general, create deviations (however slight) from the Born Rule. The non-unitarity of RTI arises naturally from a different mechanics of the underlying field behavior, not from changing the quantum theory itself to force a non-unitary process that would otherwise not occur. In strict mathematical terms, the non-unitarity arises from the fact that the action based on the Feynman propagator (which describes the field connection between emitting and absorbing currents) is non-unitary (as shown in Chapter 5).

The reader may wonder why the transactional picture has not received more attention from the academic community in view of its ability to solve the measurement problem and to provide a physical derivation of the Born Rule. There are probably two primary reasons for this: (1) Tim Maudlin's claimed refutation of TI via a thought experiment (2002) and (2) the fact that the founders of the direct-action theory, Wheeler and Feynman, abandoned it. Regarding (1), Maudlin's objection set back serious consideration of TI for over a decade, even though it was largely refuted by several authors prior to the current fully relativistic development (Berkovitz, 2002; Kastner, 2014a; Marchildon, 2006). However, we have seen in Chapter 6 (see also Kastner, 2019a) that the relativistic development, which takes into account that there can be no "slow-moving offer wave" as required to instantiate the experiment, completely nullifies the Maudlin objection, so that is no longer at issue.

Regarding (2), Wheeler and Feynman were mainly concerned with eliminating self-action infinities, and the classical form of the direct-action theory allowed them to prohibit any interaction of a source with its own field. When they realized that some degree of self-action was required at the quantum level in order to account for such phenomena as the Lamb shift, they felt that the theory no longer served the purpose that was their primary motivation, so they abandoned it. However, there is nothing wrong with the theory itself, and Davies' quantum relativistic version permits self-action as required for the Lamb shift. Since the self-action consists only of the time-symmetric photon propagator, which cannot transfer energy, it does not amount to "self-energy" of a charge as is usually assumed. It is only self-force involving virtual (off-shell) photons, and it does not convey any real energy. Thus, even if formal infinities may arise from such self-action, they correspond only to infinite force, which can be readily treated by way of renormalization that is not subject to the embarrassing idea of "throwing away an infinite amount of energy," which Feynman referred to as "a dippy process" (Feynman, 1985). This is because there is no real energy to be thrown away. What is being thrown away is a quantum form of potential energy.[2] It is well known that the zero of potential energy is arbitrary; so there is no problem in defining it away through renormalization. Thus, the "problem" of self-action is defanged in the direct-action theory, because it is not in fact self-energy. For further details on how

---

[2] By a "quantum form," I mean that it is pre-transactional and therefore applies to the quantum substratum. It does not have units of energy, since it is not acting over any spacetime distance.

the direct-action theory can resolve long-standing problems in quantum field theory, see Kastner (2015).

In any case, the founders' abandonment of the direct-action theory was not permanent: Wheeler eventually returned to promoting it in connection with a search for a theory of quantum gravity. In a paper written with D. Wesley, he said:

[The Wheeler–Feynman theory] swept the electromagnetic field from between the charged particles and replaced it with "half-retarded, half advanced direct interaction" between particle and particle. It was the high point of this work to show that the standard and well-tested force of reaction of radiation on an accelerated charge is accounted for as the sum of the direct actions on that charge by all the charges of any distant complete absorber. Such a formulation enforces global physical laws, and results in a quantitatively correct description of radiative phenomena, without assigning stress-energy to the electromagnetic field. (Wesley and Wheeler, 2003, p. 427)

Unfortunately, the fact that the founders' abandonment of the theory did not actually discredit it, and the fact that Wheeler eventually returned to it, are little known or understood. The idea that these towering geniuses abandoned their theory, even if only temporarily in Wheeler's case, has unfortunately left something of a stigma on the model. Nevertheless, many prominent researchers have explored the direct-action theory: Narlikar (1968), Pegg (1975), Tipler (1975), Jaynes (1990), and Rohrlich (1973).[3]

Perhaps another reason for the reluctance on the part of the "mainstream" to consider the transactional picture is that the field behavior in the direct-action theory is unfamiliar and counterintuitive. There are two main aspects to its counterintuitive nature: (1) the nonlocality of the basic time-symmetric (unitary) interaction and (2) the mutuality of emitters and absorbers in (non-unitary) energy transfer. Concerning (1), this is the "direct-action" aspect of the model: sources (charges) have a direct elementary connection (the time-symmetric propagator) that is not mediated in the usual way by a separate mechanical (quantized) field system. While RTI views the connection as a field (since it is represented by the time-symmetric propagator, a field construct), it is an immediate, nonlocal connection between charges that has no temporal direction. This is highly counterintuitive and may seem far-fetched to physicists who have been trained to think that "proper physics" involves a local, mediated account in which a field influence is always localized at some spacetime point as it proceeds, in "causal" fashion, from one charge to another. We might refer to this typical concept of field propagation as a "bucket brigade" account of field behavior. The direct-action

---

[3] Indeed, RTI's picture of field behavior, in which the Coulomb (time) component of the field is nonquantized (corresponding to virtual photons) while the transverse components are quantized (corresponding to real photons) is very similar to Rohrlich's model. Rohrlich stipulates the quantization conditions, while in RTI, quantization of transverse components arises from absorber response.

theory strongly violates the local "bucket brigade" picture. For physicists who are convinced that a key interpretive goal is "saving locality," a direct, unmediated connection between charges is obviously a barrier to consideration of the model.

Nevertheless, it should be noted that the direct connection (corresponding to virtual, off-shell photons) acts at a pre-spacetime level in the quantum substratum. In contrast, a real, on-shell photon is indeed transferred in a local manner, at light speed, from emitter to absorber as a spacetime process. This creates a consistent ontological account, since we already know that quantum correlations (such as EPR-type correlations) are nonlocal, whereas we retain locality at the spacetime level whenever real energy (as a real photon) is transferred. We can understand the distinction as one of information versus energy: one can have *information* transfer without involving actual *energy* transfer. Information (the direct field connection, as well as the entanglement correlations) acts in the quantum substratum, while energy transfer is a spacetime process (corresponding to the creation of a spacetime interval). The latter brings us to feature (2).

Feature (2), non-unitarity, has been hidden in the standard approach to quantum theory. This is because the latter assumes a unilateral model of field behavior that precludes the crucial mechanism of non-unitarity: the "response of the absorber" in the nonrelativistic original TI but, more accurately, the NU-interaction at the relativistic level (a more symmetric interaction in which emitter and responding absorbers participate together to create the real photon). The counterintuitive aspect attending this process is the strong mutuality of the interaction between emitters and absorbers that yields radiation (i.e., the loss of energy by a charge via photon transfer); it is not unilateral. This mutuality violates our commonsense expectations. The term "radiation" itself seems to presuppose a unilateral process: we may find ourselves picturing a sun that has rays emanating from it that continue on indefinitely. We assume that, like a pitcher throwing a baseball, a source can unilaterally "throw" a photon out and it will have a determinate trajectory regardless of whether it is "caught" (absorbed) by another charged object. This commonsense notion is what is denied in the transactional picture. But if we think about it (and as argued in Chapter 5), this unilateral picture of radiation is neither theoretically necessary nor empirically demonstrated. We don't in fact know that photons continue on indefinitely. We just assume it.

In addition, there may well be cultural influences underlying our naive commonsense notion of radiation as unilateral. In terms of the Eastern concepts of yin and yang (Figure 9.1), the above-described unilateral picture of radiation from traditional Western science is "all yang and no yin." Yang is the giving, creating, or initiating principle, while yin is the receptive, annihilating, or responding principle. Our yang-constrained concept of radiation is what limits our ability to resolve the "paradoxes" of quantum theory. It is roughly analogous to thinking that

Figure 9.1 Yin and yang.

all it takes to have a flower is for someone to "emit" a flower seed. But of course, sprinkling a seed out of a flower packet is a necessary but not sufficient condition. The seed must be received in the soil, with which it crucially interacts, and there must be suitable conditions for the flower to grow. Similarly, nobody unilaterally sells a house. There must be a buyer to receive the house, or there is no sale.

An even more striking example of the Western neglect of the yin component is evident in the lack of scientific progress for many decades in reproductive biology, due to an insistent mythology of the ability of the male sperm to unilaterally fertilize the female ovum. In 1991, Emily Martin argued that, in her words, biological science "has constructed a romance based on stereotypical male/female roles." Her study of the history of reproductive biology revealed that long-standing narratives portrayed the female ovum as a completely passive object, in which sperm unilaterally planted themselves. This had impeded open-minded investigation of exactly what really *did* happen during fertilization and delayed progress in understanding the details of the process. After many decades of this inattention to the actual details, researchers discovered quite by accident in the early 1980s that this standard narrative was highly inaccurate. In fact, they found that the egg was an active partner, sending forth chemical signals (without which sperm would get lost in dead ends) and creating structures to guide the sperm in and to enclose it. In effect, the sperm and the ovum were involved in a collaborative project; both players were crucial to its success. The findings were summarized as follows by Schatten and Schatten: "the egg is not merely a large, yolk-filled sphere into which the sperm burrows to endow new life. Rather, recent research suggests the almost heretical view that sperm and egg are mutually active partners" (Schatten and Schatten, 1984, p. 51). The view was only "heretical" because, as Martin notes, researchers were in the grip of misleading stereotypical assumptions concerning the roles of the sperm and the egg.

In each case discussed above – the growth of a flower, the house sale, and conception – the actively receiving part is the yin aspect, without which none of these processes could occur. The direct-action theory adds the yin component in the collaborative interaction between the emitter (the yang element) and the absorber (the yin element). The transferred photon is a process (energy transfer) that only occurs because *both* elements, giving/receiving, creating/destroying, or initiating/responding, are involved. And, as discussed in Chapter 5, this means that photons are not just emitted, as assumed in the usual unilateral mythology of "one way" fields in physics. They are *always* emitted and responded to/absorbed. Otherwise, there can be no radiation at all.

Though perhaps a "heretical" notion, the crucial role of the female ovum, contributing the yin aspect, turned out to be correct. Perhaps it is time to take seriously the parallel "heresy" of the direct-action picture of fields and the transactional picture, so that similar progress can be made in understanding and resolving the paradoxes of quantum theory.

References

Ackerhalt, K., and Rząźewski, J. R. (1975). "Heisenberg-picture operator perturbation theory." *Physics Review* **12**:6, 2549.

Afshar, S. S. (2005). "Violation of the principle of complementarity, and its implications." *Proceedings of SPIE* **5866**, 229–44.

Akhiezer, A. I., and Berestetskii, V. B. (1965). *Quantum Electrodynamics*. New York: Interscience.

Allison, W. W. M., and Cobb, J. H. (1980). "Relativistic charged particle identification by energy loss." *Annual Review of Nuclear and Particle Science* **30**, 253–98.

Anandan, J. (1997). "Classical and quantum physical geometry." In R. S. Cohen, M. Horne and J. Stachel (eds.), *Potentiality, Entanglement and Passion-at-a-Distance: Quantum Mechanical Studies for Abner Shimony*, vol. 2. Dordrecht: Kluwer, pp. 31–52. Preprint: http://arxiv.org/PS_cache/gr-qc/pdf/9712/9712015v1.pdf.

Arndt, M., and Zeilinger, A. (2003). "Buckeyballs and the dual-slit experiment." In J. Al-Khalili (ed.), *Quantum: A Guide for the Perplexed*. London: Weidenfeld & Nicolson.

Auyang, S. (1995). *How Is Quantum Field Theory Possible?* New York: Oxford University Press.

Bacciagaluppi, G. (2016). "The role of decoherence in quantum mechanics." In E. N. Zalta (ed.), *The Stanford Encyclopedia of Philosophy* (Fall 2016 edition). https://plato.stanford.edu/archives/fall2016/entries/qm-decoherence/.

Bacciagaluppi, G., and Crull, E. (2009). "Heisenberg (and Schrödinger, and Pauli) on hidden variables." http://philsci-archive.pitt.edu/archive/00004759/01/SHPMP_paper_07_10_09.pdf.

Barbour, J. B. (1982). "Relational concepts of space and time." *British Journal for the Philosophy of Science* **33**, 251–74.

Barnum, H. (1990). "Dieks' realistic interpretation of quantum mechanics: a comment." Preprint: philsci-archive.pitt.edu/2649/1/dieks.pdf.

Bell, J. S. (1990). "Against measurement." *Physics World*, August, 33–41.

Bennett, C. L. (1987). "Precausal quantum mechanics." *Physical Review A* **36**, 4139–48.

Berestetskii, V., Lifschitz, E., and Petaevskii, L. (1971). *Quantum Electrodynamics*. Landau and Lifshitz Course of Theoretical Physics, vol. 4. Amsterdam: Elsevier.

Berestetskii, V., Lifschitz, E., and Petaevskii, L. (2004). *Quantum Electrodynamics*, 2nd ed. Landau and Lifshitz Course of Theoretical Physics, vol. 4. Amsterdam: Elsevier.

Berkovitz, J. (2002). "On causal loops in the quantum realm." In T. Placek and J. Butterfield (eds.), *Proceedings of the NATO Advanced Research Workshop on Modality, Probability and Bell's Theorems*. Dordrecht: Kluwer, pp. 233–55.

Bethe, H. (1930). "Zur Theorie des Durchgangs schneller Korpuskularstrahlen durch Materie." *Annalen der Physik* **397**, 325–400.

Bialynicki-Birula, I. (1996). *Coherence and Quantum Optics VII*, ed. J. Eberly, L. Mandel, and E. Wolf. New York: Plenum Press, p. 313.

Bjorken, J., and Drell, S. (1965). *Relativistic Quantum Fields*. New York: McGraw-Hill.

Bohr, A., Mottelson, B., and Ulfbeck, O. (2003). "The principle underlying quantum mechanics." *Foundations of Physics* **34**, 405–17.

Bombelli, L., Lee, J., Meyer, D., and Sorkin, R. D. (1987). "Spacetime as a causal set." *Physics Review Letters* **59**, 521–24.

Bouchard, F., Harris, J., Mand, H., Bent, N., Santamato, E., Boyd, R. W., and Karimi, E. (2015). "Observation of quantum recoherence of photons by spatial propagation." *Scientific Reports 5*, https://doi.org/10.1038/srep15330.

Braddon-Mitchell, D. (2004). "How do we know it is now now?" *Analysis* **64**, 199–203.

Breitenbach, G. Schiller, S., and Mlynek, J. (1997). "Measurement of the quantum states of squeezed light." *Nature* **387**, 471.

Breuer, H.-P., and Petruccioni, F. (2000). *Relativistic Quantum Measurement and Decoherence: Lectures of a Workshop Held at the Institute Italiano per gli Studi Filosofici*. Berlin: Springer.

Broad, C. D. (1923). *Scientific Thought*. New York: Harcourt, Brace and Co.

Brown, H. (2002). *Physical Relativity*. Oxford: Oxford University Press.

Brown, H. R., and Wallace, D. (2005). "Solving the measurement problem: de Broglie–Bohm loses out to Everett." *Foundations of Physics* **35**, 517–40.

Brush, S. (1976). *The Kind of Motion We Call Heat: History of the Kinetic Theory of Gases in the Nineteenth Century*. Amsterdam: Elsevier.

Bub, J. (1997). *Interpreting the Quantum World*. Cambridge: Cambridge University Press.

Bub, J. (2017). "Why Bohr was (mostly) right." Preprint. https://arxiv.org/pdf/1711.01604.pdf.

Bub, J., Clifton, R., and Monton, B. (1997). "The bare theory has no clothes." In R. Healey and G. Hellman (eds.), *Minnesota Studies in Philosophy of Science XVII*. Minneapolis: University of Minnesota Press.

Butterfield, J. (2011). "On time *chez* Dummett." Preprint: http://philsci-archive.pitt.edu/8848/1/ChezDummettJNB.pdf. (A shorter version is forthcoming in a special issue of *European Journal of Analytical Philosophy* in honor of Michael Dummett.)

Callender, C. (2002). "Thermodynamic asymmetry in time." In E. N. Zalta (ed.), *The Stanford Encyclopedia of Philosophy* (Spring 2002 edition). http://plato.stanford.edu/archives/win2001/entries/time-thermo/.

Callender, C. (2010). "Is time an illusion?" *Scientific American* **302**:6, 58–65.

Chakravartty, A. (2011). "Scientific realism." In E. N. Zalta (ed.), *The Stanford Encyclopedia of Philosophy* (Summer 2011 edition), http://plato.stanford.edu/archives/sum2011/entries/scientific-realism/.

Chiatti, L. (1995). "The path integral and transactional interpretation." *Foundations of Physics* **25**, 481–90.

Clifton, R., and Halvorsen, H. (2000). "Are Rindler quanta real? Inequivalent concepts in quantum field theory." Preprint: http://philsci-archive.pitt.edu/73/1/rindler.pdf.

Clifton, R., and Monton, B. (1999). "Losing your marbles in wavefunction collapse theories." *British Journal of Philosophical Science* **50**:4, 697–717.

Cramer, J. G. (1980). "Generalized absorber theory and the Einstein–Podolsky–Rosen paradox." *Physical Review D* **22**, 362–76.

Cramer, J. G. (1983). "The arrow of electromagnetic time and the generalized absorber theory." *Foundations of Physics* **13**, 887–902.

Cramer, J. G. (1986). "The transactional interpretation of quantum mechanics." *Reviews of Modern Physics* **58**, 647–88.

Cramer, J. G. (1988). "An overview of the transactional interpretation." *International Journal of Theoretical Physics* **27**, 227.

Cramer, J. G. (2005). "The quantum handshake: a review of the transactional interpretation of quantum mechanics." Presented at "Time-Symmetry in Quantum Mechanics" Conference, Sydney, Australia, July 23, 2005. Available at: http://faculty .washington. edu/jcramer/PowerPoint/Sydney_20050723_a.ppt.

Davies, P. C. W. (1970). "A quantum theory of Wheeler–Feynman electrodynamics." *Proceedings of the Cambridge Philosophical Society* **68**, 751.

Davies, P. C. W. (1971). "Extension of Wheeler–Feynman quantum theory to the relativistic domain I. Scattering processes." *Journal of Physics A: General Physics* **6**, 836.

Davies, P. C. W. (1972). "Extension of Wheeler–Feynman quantum theory to the relativistic domain II. Emission processes." *Journal of Physics A: General Physics* **5**, 1025–36.

Davisson, C. J. (1928). "Are electrons waves?" *Franklin Institute Journal* **205**, 597.

de Broglie, L. (1923). *Comptes Rendues* **177**, 507–10.

de Broglie, L. (1925). "Recherches sur la théorie des quanta." *Annales de Physique*, 10th series, **3** (January–February).

Descartes, R. (1664). *Le Monde, ou Traite de la Lumiere.* Paris: Chez Michel Bobin.

Deutsch, D. (1999). "Quantum theory of probability and decisions." *Proceedings of the Royal Society of London* **A455**, 3129–37.

Devitt, M. (1991). *Realism and Truth*, 2nd ed. Oxford: Blackwell.

DeWitt, B. (1970). "Quantum mechanics and reality: could the solution to the dilemma of indeterminism be a universe in which all possible outcomes of an experiment actually occur?" *Physics Today* **23**, 30–40.

DeWitt, B. (2003). *The Global Approach to Quantum Field Theory*, vol. 2. Oxford: Oxford University Press.

Dirac, P. A. M. (1927). "The quantum theory of the emission and absorption of radiation." *Proceedings of the Royal Society, Series A* **114**, 243–65.

Dirac, P. A. M. (1938). *Proceedings of the Royal Society, Series A* **167**, 148–68.

Dolce, D. (2011). "De Broglie deterministic dice and emerging relativistic quantum mechanics." *Journal of Physics: Conference Series* **306**, 012049.

Dorato, M., and Felline, L. (2011). "Scientific explanation and scientific structuralism." In A. Bokulich and P. Bokulich (eds.), *Scientific Structuralism*. Boston Studies in Philosophy of Science. Berlin: Springer. Preprint: http://philsciarchive.pitt.edu/ 5095/ 1/Structural_Explanationsfinal.pdf.

Dowe, P. (2000). *Physical Causation*. New York: Cambridge University Press.

Dugić, M., and Jeknić-Dugić, J. (2012). "Parallel decoherence in composite quantum systems." *Pramana* **79**, 199.

Dummett, M. (1964). "Bringing about the past." *Philosophical Review* **73**:3, 338–59.

Dyson, F. (2009). "Birds and frogs." *AMS Einstein Lecture. Notices of the AMS* **56**, 212–23.

Earman, J. (1986). "Why space is not a substance (at least not to first degree)." *Pacific Philosophical Quarterly* **67**, 225–44.

Earman, J. (2008). "Reassessing the prospects for a growing block model of the universe." *International Studies in Philosophy of Science* **22**, 135–64.

Eddingtion, A. S. (1960). *The Mathematical Theory of Relativity*. Cambridge: Cambridge University Press.

Einstein. A. (1952). *Relativity and the Problem of Space: The Special and the General Theory*. 5th ed. (English trans.: 1954). Available at: www.relativitybook.com/resources/Einstein_space.html.

Einstein, A. (2010). *Sidelights on Relativity*. Whitefish, MT: Kessinger.

Einstein, A., Podolsky, B., and Rosen, N. (1935). "Can quantum-mechanical description of physical reality be considered complete?" *Physical Review* **47**:10, 777–80.

Eisberg, R., and Resnick, R. (1974). *Quantum Physics of Atoms, Molecules, Solids, Nuclei, and Particles*. New York: John Wiley & Sons.

Ellerman, D. (2015). "Why delayed choice experiments do not imply retrocausality." *Quantum Studies: Mathematics and Foundations* **2**:2, 183–99.

Elitzur, A. C., and Vaidman, L. (1993). "Quantum mechanical interaction-free measurements." *Foundations of Physics* **23**, 987–97.

Ellis, G., and Rothman, T. (2010). "Time and spacetime: the crystallizing block universe." *International Journal of Theoretical Physics* **49**, 988–1003.

Englert, F., and Brout, R. (1964). "Broken symmetry and the mass of gauge vector mesons." *Physical Review Letters* **13**, 321–23.

Everett, H. (1957). "Relative state formulation of quantum mechanics." *Reviews of Modern Physics* **29**, 454–62.

Falkenburg, B. (2010). *Particle Metaphysics*. Berlin: Springer.

Fearn, H. (2016). "A delayed choice quantum eraser explained by the transactional interpretation of quantum mechanics." *Foundations of Physics* **46**, 44–69. DOI 10.1007/s10701-015-9956-8.

Feynman, R. P. (1950). "Mathematical formulation of the quantum theory of electromagnetic interaction." *Physical Review* **80**, 440–57.

Feynman, R. P. (1966). "The development of the space-time view of quantum electrodynamics." Nobel Prize Lecture. *Physics Today* **19**:8, 31. https://doi.org/10.1063/1.3048404.

Feynman, R. P. (1985). *QED: The Strange Theory of Light and Matter*. Princeton, NJ: Princeton University Press.

Feynman, R. P. (1998). *Theory of Fundamental Processes*. Boulder, CO: Westview Press.

Feynman, R. P., and Hibbs, A. R. (1965). *Quantum Mechanics and Path Integrals*. New York: McGraw-Hill.

Feynman, R. P., Leighton, R., and Sands, M. (1964). *The Feynman Lectures on Physics*, vol. 3. New York: Addison-Wesley.

Fields, C. (2011). "Classical system boundaries cannot be determined within quantum Darwinism." *Physics Essays* **24**, 518–22.

Forrest, P. (2004). "The real but dead past: a reply to Braddon–Mitchell." *Analysis* **64**, 358–62.

Frauchiger, D., and Renner, R. (2018). "Quantum theory cannot consistently describe the use of itself." *Nature Communications* **9**, article number 3711.

Friedman, M. (1986). *Foundations of Space-Time Theories*. Princeton, NJ: Princeton University Press.

Frigg, R. (2005). "Review of Kuhlman, Lyre and Wayne (2002)." *Philosophy of Science* **72**, 511–14.

Ghirardi, G. C., Rimini, A., and Weber, T. (1986). "Unified dynamics for microscopic and macroscopic systems." *Physical Review D* **34**, 470.

Gisin, N. (2010). "The free will theorem, stochastic quantum dynamics and true becoming in relativistic quantum physics." arXiv:1002.1392v1 (quant-ph).

Gisin, N. (2017). "Collapse. What else?" arXiv:1701.08300.

Grangier, P., Aspect, A., and Vigué, J. (1985). "Quantum interference effect for two atoms radiating a single photon." *Physics Review Letters* **54**:5, 418–21.

Greaves, H. (2004). "Understanding Deutsch's probability in a deterministic multiverse." *Studies in History and Philosophy of Modern Physics* **35**, 423–56.

Grunbaum, A. (1973). *Philosophical Problems of Space and Time*. Dordrecht: Kluwer.

Gründler, G. (2015). "Remarks on Wheeler–Feynman absorber theory." Preprint: https://arxiv.org/pdf/1501.03516.pdf.

Guralnik, G. S., Hagen, C. R., and Kibble, T. W. B. (1964). "Global conservation laws and massless particles." *Physical Review Letters* **13**, 585–87.

Hacking, I. (1983). *Representing and Intervening: Introductory Topics in the Philosophy of Natural Science*. Cambridge: Cambridge University Press.

Hardy, L. (1992a). "Quantum mechanics, local realistic theories, and Lorentz-invariant realistic theories." *Physical Review Letters* **68**:20, 2981–84.

Hardy, L. (1992b). "On the existence of empty waves in quantum theory." *Physical Letters A* **167**, 11–16.

Heathwood, C. (2005). "The real price of the dead past: a reply to Forrest and to Braddon–Mitchell." *Analysis* **65**, 249–51.

Heisenberg, W. (1958). *Physics and Philosophy*. New York: Harper Row.

Heisenberg, W. (2007). *Physics and Philosophy*. Harper Perennial Modern Classics edition. New York: HarperCollins.

Hellwig, K. E., and Kraus, K. (1970). "Formal description of measurements in local quantum field theory." *Physical Review D* **1**, 566–71.

Henson, J. (2015). "How causal is quantum theory?" Talk given at New Directions in Physics Conference, Washington, DC. Sponsored by Committee on Philosophy and the Sciences, UMCP. carnap.umd.edu/philphysics/hensonslides.pptx.

Hestenes, D. (2010). "Electron time, mass and zitter." Preprint: https://core.ac.uk/display/21250651.

Hestenes, D. (2019). "Quantum mechanics of the electron particle-clock." Preprint: https://arxiv.org/pdf/1910.10478.pdf.

Higgs, P. W. (1964). "Broken symmetries and the masses of gauge bosons." *Physical Review Letters* **13**, 508–9.

Hoyle, F., and Narlikar, J. V. (1969). "The quantum mechanical response of the universe." *Annals of Physics* **54**, 207–39.

Hughes, R. I. G. (1989). *The Structure and Interpretation of Quantum Mechanics*. Cambridge, MA: Harvard University Press.

Jammer, M. (1993). *Concepts of Space: The History of Theories of Space in Physics*. New York: Dover.

Jaynes, E. T. (1990). "Probability in quantum theory." In W. H. Zurek (ed.), *Complexity, Entropy, and the Physics of Information*. Redwood City, CA: Addison-Wesley, p. 381. Revised and extended version available at: https://bayes.wustl.edu/etj/articles/prob.in.qm.pdf.

Joos, E. (1986). "Why do we observe a classical spacetime?" *Physics Letters A* **116**, 6–8.

Joos, E., and Zeh, H. D. (1985). "The emergence of classical properties through interaction with the environment." *Zeitschrift für Physik B* **59**, 223–43.

Kant, I. (1996). *Critique of Pure Reason*. Indianapolis, IN: Hackett. English translation, Werner Pluha.

Kastner, R. E. (1999). "Time-symmetrized quantum theory, counterfactuals, and 'advanced action.'" *Studies in History and Philosophy of Modern Physics* **30**, 237–59.

Kastner, R. E. (2005). "Why the Afshar experiment does not refute complementarity." *Studies in History and Philosophy of Modern Physics* **36**:4, 649–58.

Kastner, R. E. (2008). "The transactional interpretation, counterfactuals, and weak values in quantum theory." *Studies in History and Philosophy of Modern Physics* **39**, 806–18.

Kastner, R. E. (2011a). "On delayed choice and contingent absorber experiments." *ISRN Mathematical Physics* **2012**, doi:10.5402/ 2012/617291.

Kastner, R. E. (2011b). "The broken symmetry of time." *AIP Conference Proceedings* **1408**, 7–21. doi:10.1063/1.3663714.

Kastner, R. E. (2011c). "Quantum nonlocality: not eliminated by the Heisenberg picture." *Foundations of Physics* **41**, 1137–42. https://doi.org/10.1007/s10701-011-9536-5.

Kastner, R. E. (2013). "De Broglie waves as the 'bridge of becoming' between quantum theory and relativity." *Foundations of Science* **18**, 1–9. doi:10.1007/s10699-011-9273-4.

Kastner, R. E. (2014a). "Maudlin's challenge refuted: reply to Lewis." *Studies in History and Philosophy of Modern Physics* **47**, 15–20.

Kastner, R. E. (2014b). "Einselection of pointer observables: the new H-theorem?" *Studies in History and Philosophy of Modern Physics* **48**, 56–58. Preprint: http://philsci-archive.pitt.edu/10757/.

Kastner, R. E. (2015). "Haag's theorem as a reason to reconsider direct-action theories." *International Journal of Quantum Foundations* **1**:2, 56–64. arXiv:1502.03814.

Kastner, R. E. (2016a). "Antimatter in the direct-action theory of fields." *Quanta* **5**:1, 12–18. arXiv:1509.06040.

Kastner, R. E. (2016b). "Beyond complementarity." In R. E. Kastner, J. Jeknic-Dugic, and G. Jaroszkiewicz (eds.), *Quantum Structural Studies*. Singapore: World Scientific.

Kastner, R. E. (2016c). "The Born Rule and free will." In C. DeRonde et al. (eds.), *Probing the Meaning of Quantum Mechanics: Superposition, Dynamics, Semantics and Identity*. Singapore: World Scientific, pp. 231–43. Preprint: http://philsci-archive.pitt.edu/11893.

Kastner, R. E. (2017). "On quantum collapse as a basis for the second law of thermodynamics." *Entropy 2017*, **19**:3, 106. https://doi.org/10.3390/e19030106.

Kastner, R. E. (2018). "On the status of the measurement problem: recalling the relativistic transactional interpretation." *International Journal of Quantum Foundations* **4**:1, 128–41.

Kastner, R. E. (2019a). "The relativistic transactional interpretation: immune to the Maulin challenge." In C. de Ronde et al. (eds.), *Probing the Meaning of Quantum Mechanics: Information, Contextuality, Relationalism and Entanglement*. Singapore: World Scientific. arXiv:1610.04609.

Kastner, R. E. (2019b). "The delayed choice quantum eraser neither erases nor delays." *Foundations of Physics* **49**, 717. Preprint: https://arxiv.org/abs/1905.03137.

Kastner, R. E. (2019c). *Adventures in Quantumland: Exploring Our Unseen Reality*. Singapore: World Scientific.

Kastner, R. E. (2020a). "Decoherence in the Transactional Interpretation." *International Journal of Quantum Foundations* **6**:2, 24–39.

Kastner, R. E. (2020b). "Unitary-only quantum theory cannot consistently describe the use of itself: on the Frauchiger–Renner paradox." *Foundations of Physics* **50**, 441–56. https://doi.org/10.1007/s10701-020-00336-6.

Kastner, R. E., and Cramer, J. G. (2018). "Quantifying absorption in the Transactional Interpretation." *International Journal of Quantum Foundations* **4**:3, 210–22.

Kastner, R. E., Kauffman, S., and Epperson, M. (2018). "Taking Heisenberg's potentia seriously." *International Journal of Quantum Foundations* **4**:2, 158–72.

Kent, A. (2010). "One world versus many: the inadequacy of Everettian accounts of evolution, probability, and scientific confirmation." In S. Saunders, J. Barrett, A. Kent, and D. Wallace (eds.), *Many Worlds? Everett, Quantum Theory and Reality*. Oxford: Oxford University Press.

Kiefer, C., and Joos, E. (1999). "Decoherence: concepts and examples." In P. Blanchard and A. Jadczyk (eds.), *Quantum Future from Volta and Como to the Present and Beyond*. Lecture Notes in Physics, vol. 517. Berlin: Springer.

Kim, Y.-H., Yu, R., Kulik, S. P., Shih, Y. H., and Scully, M. (2000). "A delayed choice quantum eraser." *Physical Review Letters* **84**, 1–5.

Kokorowski, D., Cronin, A. D., Roberts, T. D., and Pritchard, D. E. (2000). "From single to multiple-photon decoherence in an atom interferometer." *Physics Review Letters* **86**:11. DOI: 10.1103/PhysRevLett.86.2191. Preprint: arxiv:quant-ph/0009044.

Kondo, J. (1964). "Resistance minimum in dilute magnetic alloys." *Progress of Theoretical Physics* **32**, 37.

Konopinski, E. J. (1980). *Electromagnetic Fields and Relativistic Particles*. New York: McGraw-Hill.

Knuth, K., and Bahreyni, N. (2012). "A potential foundation for emergent space-time." *Journal of Mathematical Physics* **55**, 112501.

Ladyman, J. (2009). "Structural realism." In E. N. Zalta (ed.), *The Stanford Encyclopedia of Philosophy* (Summer 2009 edition). http://plato.stanford.edu/archives/ sum2009/ entries/structural-realism/.

Lamb, W., and Retherford, R. (1947). "Fine structure of the hydrogen atom by a microwave method." *Physical Review* **72**:3, 241–43.

Landau, L. D., and Peierls, R. (1931). "Erweiterung des Unbestimmtheitsprinzips für die relativistische Quantentheorie." *Zeitschrift für Physik* **69**, 56.

Lewis, D. (1986). *On the Plurality of Worlds*. Oxford: Blackwell.

Mach, E. (1914). *The Analysis of Sensations, and the Relation of the Physical to the Psychical*. Open Court.

MacKinnon, E. (2005). "Generating ontology: from quantum mechanics to quantum field theory." Preprint: http://philsci-archive.pitt.edu/2467/1/Ontology.pdf.

Mandel, L. (1966). "Configuration-space photon number operators in quantum optics." *Physical Review* **144**, 1071.

Mandl, F., and Shaw, G. (1990). *Quantum Field Theory*. New York: John Wiley & Sons.

Marchildon, L. (2006). "Causal loops and collapse in the transactional interpretation of quantum mechanics." *Physics Essays* **19**, 422.

Marchildon, L. (2008). "On relativistic element of reality." *Foundations of Physics* **38**, 804–17.

Martin, E. (1991). "The egg and the sperm: how science has constructed a romance based on stereotypical male/female roles." *Signs* **16**, 485–501. https://web.stanford.edu/ ~eckert/PDF/Martin1991.pdf.

Maudlin, T. (1995). "Why Bohm's theory solves the measurement problem." *Philosophy of Science* **62**, 479–83.

Maudlin, T. (2002). *Quantum Nonlocality and Relativity: Metaphysical Intimations of Modern Physics*, 2nd ed. Oxford: Wiley-Blackwell.

McMullin, E. (1984). "A case for scientific realism." In J. Leplin (ed.), *Scientific Realism*. Berkeley: University of California Press.

McTaggart, J. E. (1908). "The unreality of time." *Mind: A Quarterly Review of Psychology and Philosophy* **17**, 456–73.

Mermin, D. (1989). "What's wrong with this pillow?" *Physics Today* **42**:4, 9.

Mueller, T. M. (2015). "The Boussinesq debate: reversibility, instability, and free will." *Science in Context* **28**:4, 613–35.

Narlikar, J. (1968). "On the general correspondence between field theories and the theories of direct particle interaction." *Mathematical Proceedings of the Cambridge Philosophical Society* **64**, 1071.

Neihardt, J. G. (1972). *Black Elk Speaks*. New York: Washington Square Press.

Neumaier, A. (2016). A physicist's FAQ. www.mat.univie.ac.at/~neum/physfaq/topics/position.html.

Norton, J. (2010). "Time really passes." *Humana Mente: Journal of Philosophical Studies* **13**, 23–34.

Omnès, R. (1997). "General theory of the decoherence effect in quantum mechanics." *Physics Review A* **56**, 3383–94.

Orozco, L. (2002). "A double-slit quantum eraser experiment." http://grad.physics.sunysb.edu/~amarch/.

Pauli, W. (1928). *Festschrift zum 60sten Geburtstag A. Sommerfelds*. Leipzig: Hirzel, p. 30.

Pearle, P. (1997). 'True collapse and false collapse." In Da Hsuan Feng and Bei Lok Hu (eds.), *Quantum Classical Correspondence: Proceedings of the 4th Drexel Symposium on Quantum Nonintegrability, Philadelphia, PA, USA, September 8–11, 1994*. Cambridge, MA: International Press, pp. 51–68.

Pegg, D. T. (1975). "Absorber theory of radiation." *Reports on Progress in Physics* **38**, 1339.

Penrose, R. (1989). *The Emperor's New Mind: Concerning Computers, Minds, and Laws of Physics*. Oxford: Oxford University Press.

Petersen, A. (1963). "The philosophy of Niels Bohr." *Bulletin of the Atomic Scientists* **19**:7.

Price, H. (1996). *Time's Arrow and Archimedes' Point*. Oxford: Oxford University Press.

Psillos, S. (1999). *Scientific Realism: How Science Tracks Truth*. London: Routledge.

Psillos, S. (2003). *Causation and Explanation*. Montreal: McGill-Queens University Press.

Pusey, M., Barrett, J., and Rudolph, T. (2011). "The quantum state cannot be interpreted statistically." Preprint: quant-ph/1111.3328.

Putnam, H. (1967). "Time and physical geometry." *Journal of Philosophy* **64**, 240–47. (Reprinted in Putnam's *Collected Papers*, vol. 1 [Cambridge: Cambridge University Press, 1975].)

Quine, W. (1953). *From a Logical Point of View*. Boston: Harvard University Press.

Redhead, M. (1995). "More ado about nothing." *Foundations of Physics* **25**, 123–37.

Reichenbach, H. (1953). "La signification philosophique du dualism ondes-corpuscles." In Andre George (ed.), *Louis de Broglie, Physicien et Penseur*. Paris: Michel Albin, p. 133.

Reichenbach, H. (1958). *The Philosophy of Space and Time*. Trans. M. Reichenbach and J. Freund. New York: Dover.

Rietdijk, C. W. (1966). "A rigorous proof of determinism derived from the special theory of relativity." *Philosophy of Science* **33**, 341–44.

Rohrlich, F. (1973). "The electron: development of the first elementary particle theory." In J. Mehra (ed.), *The Physicist's Conception of Nature*. Dordrecht: D. Reidel, pp. 331–69.

Rovelli, C. (1996). "Relational quantum mechanics." *International Journal of Theoretical Physics* **35**:8, 1637–78.

Russell, B. (1913). "On the notion of cause." *Proceedings of the Aristotelian Society* **13**, 1–26.

Russell, B. (1948). *Human Knowledge*. New York: Simon and Schuster.

Russell, B. (1959). *The Problems of Philosophy*. Oxford: Oxford University Press.

Saari, P. (2011). "Photon localization." In S. Lyagushyn (ed.), *Quantum Optics and Laser Experiments*. Rijeka: InTech Open, p. 49.

Sakurai, J. J. (1973). *Advanced Quantum Mechanics*, 4th ed. New York: Addison-Wesley.

Sakurai, J. J. (1984). *Modern Quantum Mechanics*. New York: Addison-Wesley.

Salmon, W. (1984). *Scientific Explanation and the Causal Structure of the World*. Princeton, NJ: Princeton University Press.

Salmon, W. (1997). "Causality and explanation: a reply to two critiques." *Philosophy of Science* **64**:3, 461–77.

Savitt, S. (2008). "Being and becoming in modern physics." In E. N. Zalta (ed.), *The Stanford Encyclopedia of Philosophy* (Winter 2008 edition). http://plato.stanford.edu/archives/win2008/entries/spacetime-bebecome/.

Schatten, G., and Schatten, H. (1984). "The energetic egg." *Medical World News* **23** (January 23).

Schlosshauer, M., and Fine, A. (2003). "On Zurek's derivation of the Born Rule." *Foundations of Physics* **35**:2, 197–213.

Schrödinger, E. (1935). "The present situation in quantum mechanics." *Proceedings of the American Philosophical Society* **124**, 323–38.

Seiberg, N. (2007). "Emergent spacetime." In D. Gross, M. Henneaux, and A. Sevrin (eds.), *The Quantum Structure of Space and Time: Proceedings of the 23rd Solvay Conference*. Singapore: World Scientific, pp. 163–213.

Seibt, Johanna. (2020). "Process philosophy." In E. N. Zalta (ed.), *The Stanford Encyclopedia of Philosophy* (Summer 2020 edition). https://plato.stanford.edu/archives/sum2020/entries/process-philosophy/.

Shimony, A. (2009). "Bell's theorem." In E. N. Zalta (ed.), *Stanford Encyclopedia of Philosophy* (Summer 2009 edition). https://plato.stanford.edu/archives/sum2009/entries/bell-theorem/.

Silberstein, M., Stuckey, W. M., and Cifone, M. (2008). "Why quantum mechanics favors adynamical and acausal interpretations such as relational blockworld over backwardly causal and time-symmetric rivals." *Studies in History and Philosophy of Modern Physics* **39**, 736–51.

Sklar, L. (1974). *Space, Time and Spacetime*. Berkeley: University of California Press.

Sklar, L. (2015). "Philosophy of statistical mechanics." In E. N. Zalta (ed.), *The Stanford Encyclopedia of Philosophy* (Fall 2015 edition). http://plato.stanford.edu/archives/fall2015/entries/statphys-statmech/.

Sorkin, R. D. (2003). "Causal sets: discrete gravity (notes for the Valdivia Summer School)." In A. Gomberoff and D. Marolf (eds.), *Proceedings of the Valdivia Summer School*. Preprint version.

Spekkens, R. (2007). "Evidence for the epistemic view of quantum states: a toy theory." *Physical Review A* **75**, 032110.

Stapp, H. (2011). "Retrocausal effects as a consequence of orthodox quantum mechanics refined to accommodate the principle of sufficient reason." Preprint: http://arxiv.org/abs/1105.2053.

Stein, H. (1968). "On Einstein–Minkowski space-time." *The Journal of Philosophy* **65**, 5–23.

Stein, H. (1991). "On relativity theory and openness of the future." *Philosophy of Science* **58**, 147–67.

Stewart, I., and Golubitsky, M. (1992). *Fearful Symmetry: Is God a Geometer?* Oxford: Blackwell.

Sutherland, R. (2008). "Causally symmetric Bohm model." *Studies in History and Philosophy of Modern Physics* **39**, 782–805.

Teller, P. (1997). *An Interpretive Introduction to Quantum Field Theory*. Princeton, NJ: Princeton University Press.

Teller, P. (2002). "So what is the quantum field?" In M. Kuhlman, H. Lyre, and A. Wayne (eds.), *Ontological Aspects of Quantum Field Theory*. Singapore: World Scientific, pp. 145–60.

Tipler, F. (1975). "Direct-action electrodynamics and magnetic monopoles." *Il Nuovo Cimento B* **28**:2, 446–52 (197508).

Tooley, M. (1997). *Time, Tense and Causation*. Oxford: Clarendon Press.

Toral, R. (2015). "Introduction to master equations." Preprint: https://ifisc.uib-csic.es/raul/CURSOS/SP/Introduction_to_master_equations.pdf.

Tumulka, R. (2006). "Collapse and relativity." In A. Bassi, D. Duerr, T. Weber, and N. Zanghi (eds.), *Quantum Mechanics: Are There Quantum Jumps?* AIP Conference Proceedings 844. College Park, MD: American Institute of Physics, pp. 340–52. Preprint: http://arxiv.org/ PS_cache/quant-ph/pdf/0602/0602208v2.pdf.

Valentini, A. (1992). "On the pilot-wave theory of classical, quantum and subquantum physics." PhD dissertation, International School for Advanced Studies.

van Fraassen, B. (1991). *Quantum Mechanics: An Empiricist View*. Oxford: Oxford University Press.

van Fraassen, B. (2004). *The Empiricist Stance*. New Haven, CT: Yale University Press.

von Neumann, J. (1932). *Mathematische Grundlagen der Quantenmechanik*. Julius Springer.

Wallace, D. (2006). "Epistemology quantized: circumstances in which we should come to believe in the Everett interpretation." *British Journal for the Philosophy of Science* **57**, 655–89.

Walsh, J., and Knuth, K. (2015). "An information physics derivation of equations of geodesic form from the influence network." Presented at the MaxEnt 2015 Conference, Bayesian Inference and Maximum Entropy Methods in Science and Engineering. https://arxiv.org/abs/1604.08112.

Weingard, R. (1972). "Relativity and the reality of past and future events." *British Journal for the Philosophy of Science* **23**, 119–21.

Wesley, D., and Wheeler, J. A. (2003). "Towards an action-at-a-distance concept of spacetime." In A. Ashtekar et al. (eds.), *Revisiting the Foundations of Relativistic Physics: Festschrift in Honor of John Stachel*, Boston Studies in the Philosophy and History of Science (Book 234). Dordrecht: Kluwer Academic Publishers, pp. 421–36.

Wesson, P., and Overduin, J. M. (2019). *Principles of Space-Time-Matter*. Singapore: World Scientific.

Wheeler, J. A. (1978). "The 'past' and the 'delayed-choice double-slit experiment.'" In A. R. Marlow (ed.), *Mathematical Foundations of Quantum Theory*. Amsterdam: Academic Press, pp. 9–48.

Wheeler, J. A. (1990). *A Journey into Gravity and Spacetime*. Scientific American Library. New York: W. H. Freeman.

Wheeler, J. A., and Feynman, R. P. (1945). "Interaction with the absorber as the mechanism of radiation." *Reviews of Modern Physics* **17**, 157–61.

Wheeler, J. A., and Feynman, R. P. (1949). "Classical electrodynamics in terms of direct interparticle action." *Reviews of Modern Physics* **21**, 425–33.

Wheeler, J. A., and Zurek, W. H. (1983). *Quantum Theory and Measurement*. Princeton Series in Physics. Princeton, NJ: Princeton University Press.

Wilson, K. G. (1971). "Feynman graph expansion for critical exponents." *Physical Review Letters* **28**, 548.

Wilson, K. G. (1974). "Confinement of quarks." *Physical Review D* **10**, 2445.

Wilson, K. G. (1975). "The renormalization group: critical phenomena and the Kondo problem." *Reviews of Modern Physics* **47**, 773–840.

Wilzbach, M., Haase, A., Shwartz, M., Heine, D., Wicker, K., Liu, X., Brenner, K.-H., Groth, S., Fernholz, T., Hessmo, B., and Schmiedmayer, J. (2006). "Detecting neutral atoms on an atom chip." *Fortschritte der Physik* **54**, 746–64. doi:10.1002/prop.200610323.

Worrall, J. (1989). "Structural realism: the best of both worlds?" *Dialectica* **43**, 99–124.

Zanardi, P. (2001). "Virtual quantum subsystems." *Physics Review Letters* **87**, 077901.

Zee, A. (2010). *Quantum Field Theory in a Nutshell*. Princeton, NJ: Princeton University Press.

Zeh, H. D. (1989). *The Physical Basis of the Direction of Time*. Berlin: Springer-Verlag.

Zeilinger, A. (1996). "On the interpretation and philosophical foundations of quantum mechanics." In U. Ketvel et al. (eds.), *Vastakohtien todellisuus. Festschrift for K. V. Laurikainen*. Helsinki: Helsinki University Press.

Zeilinger, A. (2005). "The message of the quantum." *Nature* **438**, 743. https://doi.org/10.1038/438743a.

Zurek, W. H. (2003). "Decoherence, einselection, and the quantum origins of the classical." *Reviews of Modern Physics* **75**, 715–75.

Index

Printed in the United States
by Baker & Taylor Publisher Services